Nuclear Danger and the NPT

Nonproliferation Policy Education Center

Nonproliferation Policy Education Center
Arlington, VA 22216
www.npolicy.org

Printed in the United States of America.

ISBN: 978-1-7371113-5-1

Cover images: Nuclear Nonproliferation Treaty logo from the U.S. Department of State and missiles drawn by Amanda Sokolski.

Nonproliferation Policy Education Center

The Nonproliferation Policy Education Center (NPEC), a 501(c)3 nonprofit organization, is a nonpartisan, educational organization founded in 1994 to promote a better understanding of strategic weapons proliferation issues. NPEC educates policymakers, journalists, and university professors about proliferation threats and possible new policies and measures to meet them.

For current publications of the Nonproliferation Policy Education Center, please visit www.npolicy.org.

Contents

Acknowledgments

This research was supported by generous grants from the Sarah Scaife Foundation, the Carnegie Corporation of New York, and the Smith Richardson Foundation.

Nuclear Danger and the NPT

Henry Sokolski

As the world's nations prepare for the 11th Nuclear Nonproliferation Treaty (NPT) conference to be held at the U.N. in New York in 2026, the prospects for proliferation should anchor their deliberations. With any bad luck and not so many more years, there is a distinct possibility South Korea, Japan, the UAE, Egypt, Turkey, and Iran will all have nuclear weapons. After that, Algeria, Finland, Sweden, Ukraine, and Australia will be sorely tempted to do the same. There could be more.

What will drive this charge? Several things: Increased Russian, Chinese, and North Korean nuclear threats, unrestrained forward basing of nuclear weapons beyond Russia's deployment to Belarus, waning allied confidence in the credibility of U.S. nuclear security guarantees, and an increasing acceptance of dangerous civilian nuclear activities that can bring states to the brink of bomb making.

Legally, the NPT is supposed to prevent these developments. Every nation but Israel, Pakistan, North Korea, India, and South Sudan has signed the NPT. It forbids assisting, encouraging, or inducing nonnuclear states "in any other way" from manufacturing, "otherwise acquiring," or gaining control over nuclear explosive devices.

The treaty also prohibits states that lack nuclear explosive devices from seeking assistance to acquire them or gain control over them. It also requires them to allow regular international inspections of their civilian nuclear activities and holdings to prevent their military diversion to build nuclear explosives. Finally, the treaty encourages the sharing of "peaceful" nuclear energy and requires each party to "pursue negotiations in good faith on effective measures relating to cessation of the nuclear arms race."

These treaty provisions are the supreme law in 191 nations. The question is to what extent any of these provisions will be fully realized in the next five to 15 years. This volume's commissioned research and diplomatic scenarios present a set of litmus tests to provide us with an answer. The conclusion is unsettling: Unless Western governments – and ultimately China and Russia – enforce the NPT more rigorously and in new ways, further proliferation and a frightening spike in diplomatic and military tensions in Eastern Europe, the Middle East, South West Asia or East Asia is entirely likely. What the United States and like-minded nations do now to prevent this matters. The book highlights what's to be done.

The first chapter is by Christopher Ford, a former assistant Secretary of State for International Security and Nonproliferation under former U.S. President Donald Trump. Mr. Ford details how best to dissuade states, like Iran, from withdrawing from the NPT. The model for Iran doing so is North Korea, which withdrew from the NPT in 2003. Mr. Ford's findings reflect a diplomatic simulation of Iran leaving the NPT that NPEC conducted in August 2021 before his chapter went final. Mr. Ford acted as a major senior U.S. appointee in the game. A description of the game's play and its key findings are attached in this volume's appendix. While Mr. Ford makes several sanctions suggestions, his most important recommendation is for the United States and like-minded nations to promote his recommendations without waiting to secure China or Russia's agreement.

Andrea Stricker of the Foundation for the Defense of Democracy makes a similar recommendation. The United States and like-minded nations, she argues, should place a moratorium on the spread of critical nuclear bomb making activities, such as uranium enrichment and plutonium recycling. What is immediately salient is the current U.S.-Saudi negotiations to allow the Kingdom to enrich uranium on Saudi soil and U.S. Department of Energy programs that promote the development and export of fast reactors, which were originally designed to use recycled plutonium. If the U.S. government proceeds with these endeavors, it's unclear how credible any future nuclear nonproliferation measures will be.

This brings us to the volume's next chapter, which focuses on current U.S. legal restrictions on the reprocessing of spent reactor fuel from U.S. reactors operating in China. As Eldon Greenberg, a retired attorney who has held senior legal counsel positions on the Hill and Executive Branch, explains that Beijing must meet a number of legal conditions before they are able to reprocess U.S. spent fuel. His points are hardly academic: The Pentagon has repeatedly warned that China may be recycling plutonium to fuel its fast reactors. Fast reactors are excellent sources of super weapons-grade plutonium, a critical resource needed to expand China's nuclear arsenal.

This raises the more general problem of Beijing's nuclear arms buildup and its refusal to negotiate in good faith on effective nuclear arms control measures as required by Article VI of the NPT. Both the Trump and Biden administration have pointed out that Beijing's refusal to join such talks constitutes a violation of the treaty. However, many hawks – myself included – have long argued that Article VI is little more than an aspirational political gesture that lacks any legal standing. In his chapter, Thomas Grant, an international lawyer and a former legal advisor to the U.S. Department of State, lays out the evidence and legal reasoning as to why this is dead wrong. His analysis played a major role in shaping another NPEC diplomatic simulation that focused on

China's noncompliance with Article VI. That game's play and findings can be found in this volume's appendix. The game's findings and Mr. Grant's analysis are required reading for anyone serious about keeping the NPT alive.

There are other ways the NPT could be made more relevant. I explore these methods in my chapter "Xi and Putin Are Building More Nukes: How to Compete." In it, I raise the increasingly nettlesome issue of nuclear weapons sharing and the question of how nations should interpret the NPT's ban on relinquishing "control" of these weapons in Articles I and II. The point of departure is Russian and Chinese nuclear hypocrisy. While Putin and Xi complain about America's 100 forward deployed nuclear weapons in NATO and the hypothetical prospect of redeployments to South Korea and Japan, Russia is redeploying weapons in Belarus and China is rattling its neighbors by building up its arsenal. I lay out several proposals that spotlight their misbehavior while moving the ball on earnest efforts to reduce nuclear threats. These proposals should be tested in this year's NPT preparatory meetings.

One of the loudest demand signals for the redeployment of U.S. nuclear weapons comes from South Korea. Given its mature nuclear industry, the nuclear threats North Korea poses, and its uncertainty about U.S. security guarantees, it also is the most likely state other than Iran to bolt from the NPT. Why might it leave the treaty? What, if anything, should be done to deter and prepare for this prospect? Richard Lawless, a former Central Intelligence Agency operative who perhaps was most responsible for blocking South Korea's first covert effort to acquire nuclear weapons nearly 50 years ago, provides us with answers. He parses his own history to guide us on why and how South Korea might try to acquire nuclear weapons again. His closing chapter, "Seoul's Current

Nuclear Weapons Dilemma: What Does History Teach?" should put any thoughtful reader on edge.

Finally, NPEC hosted a war game late in 2022 to spotlight the distinct military challenges associated with operating peaceful nuclear power plants in war zones — the hostile occupation and targeting of these plants to achieve military objectives. Russian targeting and operation of nuclear power plants in Ukraine have arguably turned the NPT's goals of ever greater peaceful utilization of nuclear power on their head. Given global interest in building nuclear plants in other war-torn regions, the question arises, what might happen in another Ukraine-like war in which power reactors, again, become military targets? The game's after action report, "Prescription for Military Paralysis: Wartime Reactor Meltdowns," makes clear that the military benefits of targeting these plants can be quite significant.

Improving defenses of existing plants and anticipating future threats before constructing new ones in the world's trouble spots deserves consideration at both the NPT Review Conference and at the International Atomic Energy Agency. So far, neither venue has focused on these matters. If we are serious about promoting the peaceful use of nuclear energy, this needs to change.

Deterring NPT Withdrawals: Lessons for the Like-Minded

The Hon. Christopher A. Ford[1]

Synopsis:

Ever since North Korea's withdrawal from the Nuclear Nonproliferation Treaty (NPT) after being caught in violation of its provisions, diplomats have debated what might be done to deter other such instances of withdrawal and nuclear weapons development – an event that few doubt would most likely be deeply destabilizing. In NPT review cycle discussions, several governments have urged measures be taken to this end, and experts such as former International Atomic Energy Agency (IAEA) Deputy Director General Pierre Goldschmidt have made thoughtful proposals for international action. To date, however, little has been done, and the international community seems nowhere near taking effective steps to address the NPT "withdrawal problem."

This article outlines proposals that have been made for addressing this challenge and explores the divergent international perspectives that seem so far to have made collective action

1. Dr. Ford served until January 2021 as U.S. Assistant Secretary of State for International Security and Nonproliferation, and performed the duties of the Under Secretary for Arms Control and International Security; before that he served as Special Assistant to the President and Senior Director for WMD and Counterproliferation on the U.S. National Security Council staff. The opinions expressed herein are entirely his own, and do not necessarily reflect the views of anyone else. Dr. Ford is grateful to the Nonproliferation Policy Education Center (NPEC) for its sponsorship of this research paper.

impossible at the United Nations level. As an alternative, it offers an agenda for forward movement that does not rely upon U.N.-based efforts, but would instead seek to create a coalition of like-minded states dedicated to collective action on the basis of national policy choice and prioritization, to disincentivize and deter any NPT withdrawal that would threaten international peace and security. It also advocates urgent attention, within U.S. alliance networks, to bolstering the credibility and effectiveness of the security guarantees such alliances provide to states that – faced with growing regional threats from the revisionist autocracies of Russia or China – might otherwise come to feel a need to withdraw from the NPT and develop nuclear weapons for self-protection.

I. The Problem of NPT Withdrawal

It has now been more than 28 years since the Democratic People's Republic of Korea (DPRK) first announced its withdrawal from the Nuclear Nonproliferation Treaty (NPT) in March 1993, after having been caught violating that Treaty – as well as its safeguards agreement with the International Atomic Energy Agency (IAEA) – by engaging in undeclared plutonium activities. North Korea "suspended" its withdrawal from the NPT just before it became effective in June of that year, but then completed its exit departure in January 2003 after having been confronted by U.S. officials with evidence of undeclared work to enrich uranium for nuclear weapons. Since its NPT withdrawal 18 years ago, North Korea has tested nuclear weapons on multiple occasions and has developed a force of Intercontinental Ballistic Missiles (ICBMs) and other nuclear weaponry with which it threatens its neighbors and the rest of the world.

Ever since Pyongyang completed its withdrawal from the NPT after having been caught cheating for the second time, the international community has struggled with the challenge of what to do about the prospect of *further* withdrawals. And indeed, even though North Korea has undertaken a continuing series of destabilizing provocations that keep reminding the world of the baleful potential consequences of NPT withdrawal, further withdrawal has in recent years been threatened by various other governments, including Iran,[2] Turkey,[3] and Saudi Arabia.[4] Several prominent politicians in South Korea, moreover, have also called for their country to consider developing nuclear weapons in response to DPRK threats,[5] and press reports have suggested the possibility of some such debates having occurred even in Japan.[6]

---

2. *See, e.g.*, "Iran threatens to quit global nuclear treaty and build a bomb," *Arab News* (January 21, 2020), *available at* https://www.arabnews.com/node/1615751/middle-east; Bozorgmehr Sharafedin, "Iran warns Trump it might withdraw from Non-Proliferation Treaty," *Reuters* (April 24, 2018), *available at* https://www.reuters.com/article/us-iran-nuclear-rouhani/iran-warns-trump-it-might-withdraw-from-non-proliferation-treaty-idUSKBN1HV0MY.

3. *See, e.g.*, Kingston Reif, "Saudi Arabia Threatens to Seek Nuclear Weapons," *Arms Control Today* (June 2018), *available at* https://www.armscontrol.org/act/2018-06/news/saudi-arabia-threatens-seek-nuclear-weapons.

4. *See, e.g.*, "Erdogan says it's unacceptable that Turkey can't have nuclear weapons," *Reuters* (September 4, 2019), *available at* https://www.reuters.com/article/us-turkey-nuclear-erdogan/erdogan-says-its-unacceptable-that-turkey-cant-have-nuclear-weapons-idUSKCN1VP2QN.

5. *See, e.g.*, Toby Dalton, Byun Sunggee, & Lee Sang Tae, "South Korea Debates Nuclear Options," *Carnegie Endowment for International Peace* (April 27, 2016), *available at* https://carnegieendowment.org/2016/04/27/south-korea-debates-nuclear-options-pub-63455.

6. *See, e.g.*, Michael Macarthur Bosack, "Revisiting Japan's nuclear arms debate," *Japan Times* (November 28, 2019), *available at* https://www.japantimes.co.jp/opinion/2019/11/28/commentary/japan-commentary/revisiting-japans-nuclear-arms-debate/; *see also, generally,* Mark Fitzpatrick, "How Japan Could Go Nuclear," *Foreign Affairs* (October 3, 2019), *available at* https://www.foreignaffairs.com/articles/asia/2019-10-03/how-japan-could-go-nuclear.

(Even Burma is reported to have considered the possibility not long ago.[7])

The international community has tried to respond to DPRK provocations through increasingly severe U.N. Security Council sanctions imposed between 2006 and 2017,[8] as well as a range of multilateral and unilateral responses by various nations.[9] The global response specifically to the question of how to make future such withdrawals less likely, however, has been all but nonexistent.

To be sure, the United States has tried periodically to raise this question, urging attention to the issue of deterring withdrawal in the NPT review process,[10] including in multiple policy papers published on the topic.[11] Nor have U.S. diplomats been alone. In 2008, for instance, South Korea joined the Americans in calling upon all states to cease nuclear cooperation with any country withdrawing from the NPT, and for steps to be taken to ensure that withdrawing states can no longer benefit from nuclear items and material they acquired while a Party to the Treaty – such as by securing their elimination or return to their original supplier. That

---

7. *See, e.g.*, Peter Crail, "Report Alleges Secret Myanmar Nuclear work," *Arms Control Today* (undated), *available* at https://www.armscontrol.org/act/2010_07/MyanmarReport.
8. *See* Kelsey Davenport, "UN Security Council Resolutions on North Korea," *Arms Control Association* (April 2018), *available at* https://www.armscontrol.org/factsheets/UN-Security-Council-Resolutions-on-North-Korea.
9. *See, e.g.*, U.S. Department of the Treasury, "North Korea Sanctions" (undated), *available at* https://home.treasury.gov/policy-issues/financial-sanctions/sanctions-programs-and-country-information/north-korea-sanctions.
10. *See, e.g.*, Daniel Horner, "U.S. Pursues Penalty for Renouncing the NPT," *Arms Control Today* (undated), *available at* https://www.armscontrol.org/act/2013-07/news-briefs/us-pursues-penalty-renouncing-npt.
11. *See, e.g.*, U.S. Department of State, Bureau of International Security and Nonproliferation, "Article X of the Nuclear Nonproliferation Treaty: Deterring and Responding to Withdrawal by Treaty Violators" (February 2, 2007), *available at* https://2001-2009.state.gov/t/isn/rls/other/80518.htm.

joint paper also urged that the U.N. Security Council act to ensure continuity of IAEA nuclear safeguards in a country that withdraws from the NPT.[12] France has also urged that a country that violates the NPT should not be permitted to escape accountability for its noncompliance simply by withdrawing, while Germany has urged that all nuclear material and technology a withdrawing country obtained while an NPT Party be thereafter restricted to exclusively peaceful purposes.[13] For its part, Luxembourg has recommended that all of a country's pre-existing nuclear materials, equipment, technologies, and facilities be "restricted to peaceful uses only" and remain subject to IAEA safeguards after withdrawal.[14]

While the question of how to deter further withdrawals has never precisely been *dropped* from the international agenda, however, it has so far been unable to elicit enough support to permit much of anything actually to be done. To date, no concrete action has been taken along any of the lines suggested by those governments, with the sole exception of U.N. Security Council Resolution 1887 of 2009, a non-binding measure that merely "encouraged" nuclear suppliers to acquire the return of items or material in the event that a recipient withdraws from the NPT, and "urged" states to make permanent IAEA safeguards a condition of nuclear supply.[15]

---

12. Republic of Korea & U.S. Department of State, "Deterring and Responding to NPT Withdrawal by Treaty Violators" (May 9, 2008), *available at* https://2001-2009.state.gov/t/isn/rls/other/114120.htm.
13. *See* George Bunn & John B. Rhinelander, "NPT Withdrawal: Time for the Security Council to Step In," *Arms Control Today* (undated), *available at* https://www.armscontrol.org/act/2005-05/features/npt-withdrawal-time-security-council-step.
14. *See* Pierre Goldschmidt, "Dealing Preventively with NPT Withdrawal," *Nonproliferation Policy Education Center* (January 23, 2020), at 11 *available at* http://npolicy.org/event_file/Pierre_Goldschmidt_Dealing_Preventively_with_NPT_Withdrawal_140220_1509.pdf.
15. U.N. Security Council Resolution 1887, S/RES/1887 (September 24, 2009), at op. ¶¶ 18 & 20, *available at* https://digitallibrary.un.org/record/665529?ln=en#record-files-collapse-header.

Somewhat embarrassingly, in fact, it has never even been quite clear whether NPT States Party actually agree upon whether or not the DPRK fully withdrew from the Treaty in the first place. This led, for example, to the dispiriting diplomatic charade in which North Korea's name plate was kept behind the chair's desk at NPT conferences rather than being either displayed (which would signal that the DPRK remained a State Party) or discarded (which would signal the legal efficacy of its 2003 withdrawal).[16] With even *that* point apparently being too contentious for the international community to handle, it is perhaps not surprising that actually addressing the broader withdrawal question has proven too much.

From the very first moment that China refused to permit the U.N. Security Council to consider restraining North Korea from withdrawing from the NPT,[17] therefore, the "withdrawal issue" has languished.

II. The Goldschmidt Proposals

This failure to act on the problem of deterring NPT withdrawal has certainly not been the result of a lack of ideas about what to do. In addition to the abovementioned proposals made in the 2000s by several NPT States Party, for instance, former IAEA Deputy Director General for Safeguards Pierre Goldschmidt – who had been running the IAEA's safeguards work when North Korea

---

16. According to the Arms Control Association, however, "[a]lthough the legality of North Korea's process of withdrawal remains in question, subsequent calls by the U.N. and the IAEA for Pyongyang to return to the NPT demonstrate a recognition that it is currently outside the treaty." Arms Control Association, "Timeline of the Nuclear Nonproliferation Treaty (NPT)" (March 2020), *available at* https://www.armscontrol.org/factsheets/Timeline-of-the-Treaty-on-the-Non-Proliferation-of-Nuclear-Weapons-NPT.
17. *See* Bunn & Rhinelander, *supra*.

withdrew from the Treaty – published an important examination of the question in January 2020.[18]

Goldschmidt acknowledged that because the NPT contains an explicit withdrawal clause in its Article X,[19] it was "impossible" to deny the right of States Party to withdraw. He argued, however, that it was *not* impossible "to put in place appropriate *preventive measures* to dissuade withdrawal from the Treaty."[20] Specifically, he urged the adoption by the U.N. Security Council of "a legally binding *generic* resolution" providing the IAEA expanded verification rights when a state is found to be in noncompliance with safeguards, ensuring that all sensitive fuel cycle facilities in a withdrawing state remain covered by "irreversible" safeguards agreements, and mandating that previously safeguarded materials and facilities not be used for weapons production after withdrawal.[21]

---

18. Goldschmidt, *supra*.

19. Treaty on the Non-Proliferation of Nuclear Weapons (opened for signature June 12, 1968; entered into force March 5, 1980) [hereinafter "NPT"], at Art. X(1) ("Each Party shall in exercising its national sovereignty have the right to withdraw from the Treaty if it decides that extraordinary events, related to the subject matter of this Treaty, have jeopardized the supreme national interests of its country. It shall give notice of such withdrawal to all other parties to the Treaty and to the United Nations Security Council three months in advance. Such notice shall include a statement of the extraordinary events it regards as having jeopardized its supreme interests."), *available at* https://www.un.org/disarmament/wmd/nuclear/npt/text/.

20. Goldschmidt, *supra*, at 1.

21. *Id.* at 5-6. Such safeguards would take the form of an INFCIRC/66-type safeguards agreement, which – unlike IAEA Comprehensive Safeguards Agreements (CSAs) modeled on INFCIRC/153 – do not depend upon the country in question being a State Party to the NPT. *See id.* at 11-12; *compare* IAEA, "the Agency's Safeguards System," INFCIRC/66/Rev.2 (September 16, 1968), *available at* https://www.iaea.org/sites/default/files/publications/documents/infcircs/1965/infcirc66r2.pdf *with* IAEA, "The Structure and Content of Agreements Between the Agency and States Required in Connection with the Treaty on the Non-Proliferation of Nuclear Weapons," INFCIRC/153 (corrected), *available at*

In Goldschmidt's view, moreover, such a "generic and legally binding resolution" would also define any announcement of NPT withdrawal as "a threat to international peace and security" under Article 39 of the United Nations Charter, thus teeing up subsequent legally-binding Council action to impose specific consequences as soon as such withdrawal became effective. (In a variation on this theme, he also suggested an alternative approach whereby the Council would decide in advance that any notice of withdrawal would be deemed a threat to international peace and security if the country in question had first been found by the IAEA to be in violation of its safeguards obligations.[22]) Goldschmidt also recommended that the Nuclear Suppliers Group (NSG) adopt the requirement of "irreversible safeguards" as "an export condition before making any nuclear-related transfer to a state."[23]

III. Withdrawal and Its Discontents

Given the attention given to the question of deterring and responding to NPT withdrawal over the years and yet the lack of progress on such matters, even as North Korea has provided the world with a clear lesson about the potential problems that such withdrawal can produce, it is worth exploring why the international community has not been willing to do more. The reasons are probably not hard to discern, however, though they vary considerably between several groups of key stakeholders.

---

https://www.iaea.org/sites/default/files/publications/documents/infcircs/1972/infcirc153.pdf.

22. Goldschmidt, *supra*, at 10 n.23.

23. *Id.* at 5-6 & 7-9.

A. Differing Political Perspectives on Withdrawal

The most discreditable actor in this drama, of course, is Iran, which – having been previously caught in violation both of its safeguards obligations and of Article II of the NPT[24] and having threatened to withdraw if the international community does not give it what are in effect extortion payments[25] – naturally has an interest in preventing anything being done to make its own potential future withdrawal more difficult. In this, unfortunately, Iran shares an interest with countries such as Saudi Arabia and Turkey, whose authoritarian leaders might contemplate such a move themselves, particularly if Iran withdraws first. Iran does not have a great many overt diplomatic allies in NPT-related debates, but they certainly stand in the way of progress.

In explaining the paralysis of the U.N. Security Council on such matters, however, one needs to look no further than Russia and China, both of whom have effectively signaled over the years that they do not actually *mind* nuclear weapons proliferation, provided

---

24. NPT, *supra*, at Art. II ("Each non-nuclear-weapon State Party to the Treaty undertakes not to receive the transfer from any transferor whatsoever of nuclear weapons or other nuclear explosive devices or of control over such weapons or explosive devices directly, or indirectly; not to manufacture or otherwise acquire nuclear weapons or other nuclear explosive devices; and not to seek or receive any assistance in the manufacture of nuclear weapons or other nuclear explosive devices."). For analyses of Iran's violations, *see, e.g.*, U.S. Department of State, *Adherence to and Compliance with Arms Control, Nonproliferation, and Disarmament Agreements and Commitments* (2020), at 35-42, *available at* https://www.state.gov/wp-content/uploads/2020/06/2020-Adherence-to-and-Compliance-with-Arms-Control-Nonproliferation-and-Disarmament-Agreements-and-Commitments-Compliance-Report.pdf; U.S. Department of State, *Adherence to and Compliance with Arms Control, Nonproliferation, and Disarmament Agreements and Commitments* (2005), at 72-80, *available at* https://2009-2017.state.gov/documents/organization/52113.pdf.

25. *See, e.g.*, Steven Erlanger, "U.S. Accuses Iran of 'Nuclear Extortion' and Vows More Sanctions," *New York Times* (July 10, 2019), *available at* https://www.nytimes.com/2019/07/10/world/europe/us-iran-nuclear-sanctions.html.

that it does not directly threaten them and that it *does* threaten the interests of the United States and its allies. During the Cold War, the Soviets drew a hard line against any possibility of proliferation to additional countries within the NATO Alliance, of course,[26] but they were willing – at least before the Sino-Soviet split – to help China develop its own weapons program,[27] though they later worked constructively with the United States to draft the NPT.[28] In more recent years, the government of Vladimir Putin seems again to have soured upon nonproliferation, undertaking an unsuccessful effort – in collaboration with pro-nuclear-weapons hardliners in the Iranian Revolutionary Guard Corps – to sabotage negotiations with Iran to rein in its nuclear program,[29] and Russian diplomats have worked to undermine IAEA safeguards around the world.[30]

---

26. *See, e.g.*, Assistant Secretary of State Christopher Ford, "Challenges of Policymaking in Responsible Nuclear Weapons Stewardship," remarks to the NATO Parliamentary Assembly (December 9, 2019) (discussing U.S.-Soviet debates in 1960s over NATO "nuclear sharing" policy), *available at* https://www.newparadigmsforum.com/p2464.
27. *See, e.g.*, John Wilson Lewis & Xue Litai, *China Builds the Bomb* (Stanford: Stanford University Press, 1988), at 52=0-53 & 60-65.
28. *See, e.g.*, Assistant Secretary of State Christopher Ford, "Opening Remarks in Celebration of the 50th Anniversary of the NPT," remarks at the United Nations (March 5, 2020), *available at* https://www.newparadigmsforum.com/p2499.
29. *See, e.g.*, Farnaz Fassihi, "Iran's Foreign Minister, in Leaked Tape, Says Revolutionary Guard Set Policies," *New York Times* (April 25, 2021) (recounting that IRGC Commander Qassim Suleimani worked with Russia to undermine nuclear negotiations with the West, and that Moscow "put all its weight" behind creating obstacles because it did not want the nuclear deal to succeed), *available at* https://www.nytimes.com/2021/04/25/world/middleeast/iran-suleimani-zarif.html. (In sharp contrast to Russia's apparent desire to *facilitate* Iranian nuclear weapons preparations, of course, official American distaste for the Iranian nuclear deal during the 2017-20 period stemmed from the belief that the agreement had not done *enough* to prevent proliferation.)
30. *See, e.g.*, Assistant Secretary of State Christopher Ford, "The Challenge and the Potential of U.S.-Russian Nonproliferation Cooperation," remarks at the James Martin Center for Nonproliferation Studies (October 22, 2018), *available at* https://www.newparadigmsforum.com/p2273.

As for China, for many years it also seems to have felt perfectly comfortable with nuclear proliferation to recipients whose possession of such weapons complicates national security planning for Beijing's adversaries. According to U.S. nuclear weapons scientists Thomas Reed and Danny Stillman, Chinese officials in the early 1980s

> apparently decided not only to tolerate, but also to actively support the proliferation of nuclear weapons within the Third World. China welcomed Pakistani nuclear scientists to Beijing and passed along information on the CHIC-4 A-bomb design to those visitors. … There is also evidence the Chinese conducted an underground nuclear test for the Pakistanis at Lop Nur on May 26, 1990, well before Pakistan's announced 1998 shots in south Asia.[31]

Even in recent years, China has continued to permit transfers of technology to Iran's missile program,[32] and – as we have seen – has worked to protect its troublesome DPRK quasi-client state from accountability at the U.N. Security Council.

The unfortunate Russian and Chinese track record of *de facto* toleration of proliferation where it principally threatens the United States and its allies would appear to carry with it the implied codicil that withdrawal from the NPT should not be made *too*

---

31. Thomas C. Reed & Danny B. Stillman, *The Nuclear Express: A Political History of the Bomb and its Proliferation* (Minneapolis: Zenith Press, 2009), at 131; *see also id.* at 252 (on the test for Pakistan at Lop Nur), 261-62 (North Korea's use of a plutonium-based "descendant" or "derivative" of China's CHIC-4 design), 269 (A.Q. Khan's provision of a CHIC-4 design to Libya), & 328-29 (China's provision of critical supplies and material to A.Q. Khan's proliferation network and use of North Korea as "the re-transfer point for the sale of nuclear and missile technology to Iran, Syria, Pakistan, Egypt, Libya, and Yemen").
32. *See, e.g.*, Assistant Secretary of State Christopher Ford, "Nonproliferation with Attitude: Counterproliferation Tools and Diplomacy in U.S. Foreign Policy," remarks at the Heritage Foundation (November 14, 2018), *available at* https://www.newparadigmsforum.com/p2298.

challenging for rogue regimes such as the DPRK or Iran, lest this makes it more difficult for Moscow or Beijing to continue to use proliferation as a tool of grand strategy as they work to restructure the U.S.-centric post-Cold War global environment to privilege their own relative power and influence. This does not mean that they necessarily *seek* proliferation in itself, and it is surely the case that both Russia and China would *fiercely* oppose NPT withdrawal and nuclear weapons development by a U.S. ally either in NATO or in the Far East. Nevertheless, it does seem to mean that Moscow and Beijing cannot be counted upon to do much to make withdrawal by Iran or some other West-vexing "rogue" regime any more difficult. Their behavior at the Security Council supports this conclusion.

Another group of states uneasy with doing more to deter NPT withdrawal can loosely be seen as centering around the so-called Non-Aligned Movement (NAM). NAM states seem to be uneasy with anything having to do with making withdrawal more unattractive, not necessarily out of any desire to *see* such withdrawal occur, but for reasons more abstract and ideological.

From a NAM point of view, the NPT is to some extent an intrinsically "unfair" instrument that divides the world into nuclear "haves" and "have-nots,"[33] and under which non-possessors of nuclear weapons are bound to foreswear them but the NPT nuclear weapons states (NWS) either unjustly do not have to give up such weaponry or have inexcusably failed to do so. Some also contend that the indefinite extension of the NPT in 1995 represented an additional layer of "bargain" under which the NWS must disarm, and without progress toward such disarmament it is felt

---

33. *See, e.g.*, Brazilian Foreign Minister Ceslo Amorim, "The Non-Proliferation Treaty is an Intrinsically Unfair Treaty," remarks to the 2010 Review Conference of the Treaty on the Non-Proliferation of Nuclear Weapons (May 3, 2010), *available at* https://mronline.org/2010/05/04/the-non-proliferation-treaty-is-an-intrinsically-unfair-treaty/.

problematic for non-possessors to remain forever bound by the Treaty's Article II obligations not to develop nuclear weapons.[34]

This NAM perspective is notably flawed, ignoring much of the NPT's own history and negotiating record,[35] confusing important questions of international law,[36] and overlooking the Treaty's nature as a "bargain" also between the nuclear *non*-possessors themselves – who, in fact, reap profound security benefits from the NPT because they "have powerful reasons not to see their neighbors or regional rivals acquire nuclear weapons, and ... would lack nuclear tools with which to deter threats from proliferators if this occurred."[37] Nevertheless, such NAM-flavored positions are apparently sincerely felt, and at the very least vociferously defended.

Through the NAM prism, making NPT withdrawal more difficult would accentuate the supposed injustice of the Treaty's structure, threatening to "lock" non-nuclear weapon states (NNWS) into constraints from which they have every reason to preserve the option of departure. To be sure, supporters of measures to deter withdrawal have often carefully qualified their position to make clear that they only wish to deter withdrawal *by a state that has*

---

34. *See, e.g.*, Daniel H. Joyner, *Interpreting the NPT* (Oxford: Oxford University Press, 2011).
35. *See, e.g.*, Christopher Ford, "Debating Disarmament," *Nonproliferation Review*, vol. 14, no. 3 (November 2007), at 402.
36. *See, e.g.*, Ambassador Norm Wulf, "Misinterpreting the NPT," book review of Joyner's *Interpreting the NPT*, in Arms *Control Today* (undated), *available at* https://www.armscontrol.org/act/2011-08/misinterpreting-npt; Christopher Ford, "Misinterpreting the NPT," remarks at the Carnegie Endowment for International Peace (September 30, 2011), *available at* https://www.newparadigmsforum.com/p1100.
37. Assistant Secretary of State Christopher Ford, "Rebutting False Disarmament Narratives in Support of NPT Diplomacy," remarks at the Heritage Foundation (May 16, 2019), *available at* https://www.newparadigmsforum.com/p2387.

been found in violation of the Treaty,[38] but this distinction appears insufficient to win significant NAM support. For the NAM, the very idea of doing *anything* to "impede" exercise of the withdrawal right specified in the NPT is apparently unacceptable.

B. The Challenge of Impeding All Withdrawal

And indeed, it must be admitted that while one neither can nor should defend a country for trying to "cure" noncompliance by withdrawing from the NPT after having been caught in a violation, there *is* a strong and principled argument against measures that would restrict the exercise of Article X withdrawal *per se*. The argument against de-privileging all NPT withdrawal has little or nothing to do with NAM-style confusions about the Treaty's structure or history, however.

To some degree, this argument lies in simple contractual fairness, inasmuch as there is at least something anomalous about retroactively changing terms on the basis of which a country signed and ratified a treaty that contains an express withdrawal provision. More fundamentally, however, there is also a substantive policy *reason* why the NPT has a withdrawal provision in the first place. Article X(1) allows withdrawal where a State Party decides that "extraordinary events, related to the subject matter of this Treaty, have jeopardized the supreme national interests of its country."[39] As North Korea has so sadly demonstrated, it is clearly possible for a country to invoke this

---

38. *See, e.g.*, U.S. Department of State, "Article X of the Nuclear Nonproliferation Treaty: Deterring and Responding to Withdrawal by Treaty Violators," *supra*; Republic of Korea & U.S. Department of State, "Deterring and Responding to NPT Withdrawal by Treaty Violators," *supra*.
39. NPT, *supra*, at Article X(1).

provision where in fact it does *not* truly face such a threat.[40] Nevertheless, one cannot say that it is *impossible* for a country ever to have a sound reason for withdrawal, and it would thus surely be wrong to try to deny the exercise of Article X(1) in all circumstances.

It may be impolite to say so in nonproliferation circles, for instance, but in some hypothetically grim future in which U.S. alliance guarantees have collapsed and an erstwhile U.S. ally such as, say, Estonia, Poland, South Korea, Japan, or even Taiwan faced an overwhelming threat of invasion and conquest by Russia or China, could one in good conscience tell that country that it must sacrifice its very existence on the altar of nonproliferation scrupulousness? In such an extreme scenario, the threatened country might well be able *truthfully* to declare – as the withdrawal provisions in the NPT require – that "extraordinary events" have "jeopardized … [its] supreme interests"[41] in ways that require the recovery of a nuclear weapons option to deter that aggression.

The International Court of Justice made clear in its 1996 advisory opinion on nuclear weapons, after all, that even the *use* of nuclear weapons – let alone their development for deterrence – would be

40. Though it had long purported to fear an unprovoked U.S. and South Korean invasion, after all, North Korea had already successfully "deterred" this extraordinarily unlikely event for decades through a combination of conventional forces and chemical and biological threats to the South Korean capital of Seoul. The potentially existential threat posed to Pyongyang by U.S. nuclear weapons forward-deployed in South Korea had also disappeared, with those devices being withdrawn in the early 1990s. *See, e.g.*, Don Oberdorfer, "U.S. Decides to Withdraw A-Weapons from South Korea," *Washington Post* (October 19, 1991), *available at* https://www.washingtonpost.com/archive/politics/1991/10/19/us-decides-to-withdraw-a-weapons-from-s-korea/3759ee3f-e9bf-4944-bfdf-2f9ea727b546/. By the time North Korea completed its withdrawal from the NPT, moreover, the United States was clearly preoccupied in the Middle East, having begun a war in Afghanistan in response to the terrorist attacks of September 11, 2001, and being then clearly preparing for yet another war in Iraq.

41. NPT, *supra*, at Art. X(1).

lawful "in an extreme circumstance of self-defence, in which the very survival of a State would be at stake."[42] If such survival were indeed potentially at stake, it would be hard to argue that NPT withdrawal under such circumstances would be entirely inappropriate.

The case for not trying to block such withdrawal, moreover, would be especially compelling if the threatened state were a rights-based, rule-of-law democracy and its adversary a brutal dictatorship, in which case the longstanding political, moral, and international legal value of self-determination by sovereign peoples[43] would also support an NNWS' withdrawal from the NPT in order to ensure self-preservation through nuclear deterrence.

---

42. International Court of Justice, "Legality of the Threat or Use of Nuclear Weapons" (advisory opinion) (July 8, 1996) [hereinafter "ICJ Nuclear Weapons Case"], at 44, ¶ 105(2)(E), *available at* https://www.icj-cij.org/public/files/case-related/95/095-19960708-ADV-01-00-EN.pdf. Under the longstanding international law concept that sovereign states are presumptively not subject to legal constraint unless some clear prohibition has been agreed by treaty or developed through customary international law, the ICJ's explicit inability to find any prohibition on nuclear weapons use under those circumstances thus necessarily means that such use would be lawful. *See, e.g.*, Permanent Court of International Justice, "The Case of the S.S. Lotus," Series A, No. 10 (September 7, 1927), at 19, https://documents.law.yale.edu/sites/default/files/ss_lotus_-_pcij_-_1927.pdf. This concept is also reflected in the U.S. *Law of War Manual*, which explains that "[t]he law of war that relates to the conduct of hostilities has generally been viewed as 'prohibitive law,' in the sense that it forbids rather than authorizes certain uses of force. For example, the lawfulness of the use of a type of weapon does not depend on [whether or not there is] authorization [for its use], but, on the contrary, on whether the weapon is prohibited."). U.S. Department of Defense, *Law of War Manual* (June 2015, updated December 2016), at 14, § 1.3.3.1, available *at* https://dod.defense.gov/Portals/1/Documents/pubs/DoD%20Law%20of%20War%20Manual%20-%20June%202015%20Updated%20Dec%202016.pdf?ver=2016-12-13-172036-190.

43. *See, e.g.*, Charter of the United Nations (opened for signature June 26, 1945, entered into force October 24, 1945), at Arts. 1(2) & 55 (emphasizing importance of respect for the principle of self-determination of peoples), *available at* https://en.wikipedia.org/wiki/Charter_of_the_United_Nations;

From the perspective of international peace and security, therefore, the honest observer must concede that not all imaginable NPT withdrawals are, as they were, created equal. This makes clear that impeding NPT withdrawal on a *per se* basis – that is, in *all* cases – would be inappropriate.

C. The Importance of Deterring (Almost) All Withdrawal

Nevertheless, one should not overread this caveat. The circumstances in which withdrawal would be justified would surely be *exceedingly* rare, only arising in an existentially grave situation that corresponds to *none* of the actual or threatened cases of withdrawal that the world has actually seen. No country has yet made a serious case that it faces such circumstances, and unless and until one truly can, we should be working very hard to deter and penalize NPT withdrawal.

Even in such a justifiable case, moreover, one should not pretend that the arrival of an additional "player" in the global nuclear weapons "game" would not intrinsically raise the risks of nuclear weapons use through a failure of deterrence, escalation, miscalculation, or accident. Some such risks may be worth paying if the alternative is the subjugation of a free people to tyranny, but such scenarios are certainly to be avoided if possible. (More on that below, with regard to the importance of ensuring that no U.S. ally ever comes to feel such a need for indigenous weaponization.)

Given the risks that would increase as a result of any proliferation, therefore, there are dangers in *any* NPT withdrawal, and in essentially any case other than the sort of existential extremity

Universal Declaration of Human Rights (December 10, 1948), at Art. 21(1) (noting that all people have the right to participate in the government of their country, directly or through freely-chosen representatives), *available at* https://www.un.org/en/about-us/universal-declaration-of-human-rights.

described above, the international community continues to have strong reasons to make withdrawal as difficult as possible. And it is not hard to envision examples.

Most obviously, of course, it will remain very important to deter withdrawal by a country discovered to be in violation of the Treaty (as occurred with North Korea). Such a state, after all, would not only have shown itself to be on a path toward nuclear weaponry – with all that this portends for international peace and security – but would also, through its violation, have demonstrated a degree of dishonesty and contempt for international law that would undermine the credibility of any protestation that its "supreme national interests" had truly been threatened.[44]

Withdrawal by a country whose hegemonic regional ambitions already present threats to its neighbors – threats that nuclear weaponry would naturally hugely exacerbate (as with Iran) – would also clearly present a threat to international peace and security, and should be deterred. The same might also be said for withdrawal by: (a) a country that seemed interested in nuclear weapons merely as a way of increasing its status and influence in a bid for regional influence and global status (*e.g.*, Turkey); (b) a state acting out of some weird sense of political pique or ideological principle (*e.g.*, the allegation that the NPT is in some fashion "unfair"); (c) a country seeking to use that move as a bargaining tool with which to coerce sanctions relief or other

---

44. A country willing to violate the NPT, one might also think, might be unlikely to comply with international legal requirements that any *use* of the nuclear weapons it would thereby acquire must conform to international humanitarian law (*e.g.*, the principles of military necessity, proportionality, distinction, and humanity). In its 1996 advisory opinion on nuclear weapons, the International Court of Justice made clear that any threat or use of nuclear weapons needed to be compatible with the requirements of the international law applicable in armed conflict, "particularly those of the principles and rules of international humanitarian law, as well as with specific obligations under treaties and other undertakings which expressly deal with nuclear weapons." ICJ Nuclear Weapons Case, *supra*, at 44, ¶ 105(2)(C) & (D).

benefits (*e.g.*, Iran); and (d) a country that sought simply to make more viable some future weaponization "option" for which circumstances of existential threat did not give it a compelling need. (Nor, surely, are these six examples exhaustive. It is a safe presumption that *any* withdrawal not clearly in response to an existential threat that cannot be met by other means should be deterred.)

Despite my initial caveat about the theoretical possibility of a justified withdrawal, therefore, Pierre Goldschmidt is clearly correct to see it as important to deter NPT withdrawal in *almost* every imaginable case. Nevertheless, the various cross-cutting State Party interests we have outlined above help explain why – despite thoughtful entreaties by Goldschmidt and others – the political stars have not yet aligned to permit general support even for such reasonable measures to disincentivize withdrawal by violators. To be sure, international stakeholders have been able to come together in response to *specific* circumstances of NPT withdrawal after the fact – though even here, the international community's collective action problems have not generally permitted such action to be taken quickly enough to make much difference.[45] Nevertheless, there seems little likelihood of agreement upon anything like the kind of pre-established "generic" approach through multilateral mechanisms at the United Nations and IAEA that Goldschmidt advocates.

---

45. *See generally, e.g.*, Special Assistant to the President Christopher Ford, "Challenges of the Nonproliferation 'OODA Loop,'" remarks to the NPEC Policy Fellows, Washington, D.C. (September 7, 2017) (expressing concern that collective action problems preclude multilateral fora such as the United Nations from taking enough international action, early enough, to meet future proliferation threats), *available at* https://www.newparadigmsforum.com/p2071.

IV. An Agenda for the Like-Minded

Thankfully, however, this does *not* mean that there is nothing that can be done to help meet this challenge by countries of goodwill whose governments are serious about nonproliferation. Lack of agreement at the United Nations level need not preclude responses by coalitions of states with the good sense both to prize nonproliferation and to see threats to international peace and security for what they are.

One place to start, in this respect, would be for like-minded governments to act together to salvage something of Goldschmidt's withdrawal-ameliorative agenda, as follows:

1) Such states should agree among themselves – in advance, and as a matter of national policy, reinforced by collective public declarations of mutually reinforcing national commitment – to act resolutely against themselves (whether or not the United Nations does) against any state that withdraws from the NPT under circumstances that create a threat to international peace and security.

2) States serious about fighting proliferation threats should establish in their national legislation, authorities for the imposition of mandatory sanctions triggered by a country's withdrawal from the NPT after having been either (a) found by the IAEA or by the U.N. Security Council to be in violation of IAEA safeguards or Article II of the NPT, or (b) found to be in such violation by appropriate national authorities in the state having such legislation. Even in the absence of a finding of violation, moreover, this sanctions legislation would also be triggered where national authorities deemed the circumstances to be such in which another state's NPT withdrawal presented a threat to the

country's national security or to international peace and security more generally.

(It is worth remembering that the IAEA is not empowered to make direct findings of NPT violations; such determinations are inherently left to individual national authorities, either acting individually or speaking together in and through fora such as the United Nations. It would be a dereliction of duty not to exercise such judgment, and states serious about nonproliferation should not be squeamish about doing so – even acting entirely on their own, if need be. A state unwilling to speak out against violations of a treaty signals that it does not actually value that instrument. Those who value the NPT must not remain silent when it is violated.)

3) The United States and its allies should establish agreed public understandings within their alliance networks, in advance, that withdrawal from the NPT by a country that has violated that Treaty or that has expressed hostility toward one or more members of such an alliance inherently presents a threat to collective security, and that such withdrawal could provide a legitimate basis for requesting alliance assistance (*e.g.*, through NATO's Article 5) as that threat develops.

4) All states serious about nonproliferation and deterring destabilizing NPT withdrawals should press for the improvement of IAEA safeguards, including universalization of the Additional Protocol, in order to maximize the chance that a violation of safeguards or of the NPT itself would be detected early enough to permit potential remedial action before nuclear weapons threats emerge. Taking inspiration from U.N. Security Council

Resolution 1887, they should also press for all NNWS to negotiate provisions with the IAEA pursuant to which indefinitely continuing INFCIRC/66-type safeguards would supersede Comprehensive Safeguards Agreements under INFCIRC/153[46] in the event of NPT withdrawal. States might even make such "backup" agreements a condition of nuclear supply and an accepted "best practices" standard for safeguards implementation. (Such steps would not affect the availability of withdrawal, but effective verification measures and continued monitoring might have some effect in deterring violations, or at least in maximizing the time available in which to try to overcome collective action problems in mobilizing an international response.)

5) As already called for in Resolution 1887, moreover, like-minded countries serious about nonproliferation should insist that "disgorgement" provisions be included in nuclear cooperation agreements, pursuant to which items, technology, or material provided under such agreements would have to be returned to their country of origin (or otherwise appropriately disposed of) in the event of NPT withdrawal by their recipient. (A country's refusal to agree to such a proviso, of course, would be interpreted as a sign of potential nuclear weapons intentions – and presumably, a compelling reason not to supply it with nuclear technology or material in the first place.) Adherence to the IAEA Additional Protocol should also be made a firm condition of nuclear supply by all technology possessor states – a requirement presently insisted upon only by the United States and Japan, to the shame and discredit of China, France, Russia, and South Korea.

---

46. *See supra*, note 21.

Having like-minded nonproliferation-responsible states adopt such measures as collective national policy priorities would not be as useful as having such things be done via legally-binding Security Council resolutions, of course. Nevertheless, moves in this direction by coalitions of like-minded governments would still represent important steps beyond current practice. They would help, at least in some degree, to make destabilizing and provocative North Korean-style withdrawals – *e.g.*, by Iran – less attractive and therefore presumably less likely.

There may also be more that a coalition of sensible like-minded states can do with regard to the heading of any need for a country's potentially *justified* NPT withdrawal – *e.g.*, by a democratic government the continued existence of which is threatened by a powerful menacing autocracy in a context in which that democracy can no longer credibly rely upon security guarantees given it by others. A constructive agenda in this respect could have several elements.

First and foremost – and informed by a clear understanding that the strength and credibility of the security guarantees provided by U.S.-led alliance networks has for many decades helped obviate any need for nuclear weapons development by NNWS within those alliances[47] – it must be a key priority to shore up and strengthen America's alliances. It is essential to ensure that despite severe and growing threats from Russia and China, no U.S. ally will ever feel that it has no choice *other* than to resort to NPT withdrawal and nuclear weapons development.

This is not a challenge for the United States alone, of course, since in order for these alliances to continue to provide credible security

---

47. *See, e.g.*, Assistant Secretary of State Christopher Ford, "U.S. Strength and Alliance Networks: The World's Most Successful Nonproliferation Tools," remarks at the Capitol Hill Club (April 18, 2019), *available at* https://www.newparadigmsforum.com/p2374.

guarantees to potentially threatened partners within them, *all* allies will have much work to do. Shirking on burden-sharing and mutually-supporting defense spending and modernization within an alliance framework, for example, must be seen as grossly irresponsible, and as directly contributing to making NPT withdrawal and subsequent nuclear weapons proliferation more likely.

Burden-sharing questions are commonly depicted as disputes between the United States and individual allies attempting to shirk their responsibilities, but they are in fact problems *between* such allies as well. A country that refuses to pay its fair share of alliance costs and live up to collective defense commitments, after all, is an ally that disregards the security of its neighbors within that alliance – thus making such neighbors less secure, and contributing to proliferation pressures.

Failure to invest in the continued efficacy of collective nuclear deterrence strategies – *e.g.*, the refusal of NATO Allies to recapitalize dual-capable aircraft as legacy platforms reach obsolescence, or to permit U.S. deployments of nuclear weaponry needed as part of the Atlantic Alliance deterrent – must also be seen as grossly irresponsible, and must be avoided. For many decades, U.S. "extended" nuclear deterrence in Europe and the Indo-Pacific has been an essential element in ensuring that regional allies do not face regional threats that can be met only by their own indigenous development of nuclear weaponry. Policies that undermine alliance deterrence strategies, therefore, make NPT withdrawal and proliferation more likely.

America's allies thus have much work to do in shoring up these alliance networks. That said, however, the biggest burden here still falls upon Washington, and it is absolutely essential that U.S. leaders remain strongly committed to preserving the credibility and effectiveness of the security guarantees – including the "extended" nuclear deterrence guarantees – that America's alliances provide to countries that might otherwise feel the need to resort to more autonomous varieties of nuclear deterrence. Ensuring the solidity of these guarantees will have both nuclear and conventional force posture implications, as well as requiring that the United States credibly signal not just its *capability* but also its *willingness* to come to allies' aid when needed.

In the nuclear realm, it will be important to retain U.S. capabilities – or build them where necessary – that provide credible potential responses to the growing threats presented by Russian and Chinese strategic and non-strategic nuclear (or dual-capable) delivery systems. The lower-yield W76-2 warhead recently deployed on some U.S. ballistic missile submarines,[48] for instance, should be retained, and the United States should move rapidly to replace the nuclear-armed submarine-launched cruise missile (SLCM) capability that was unilaterally scrapped by the Obama Administration in 2010 to the consternation of U.S. allies in the Far East threatened by North Korean and Chinese military power.[49] If the U.S. alliance guarantees that help answer the existential security requirements of America's allies are to be maintained over time and continue to obviate any need for indigenous allied nuclear

---

48. *See, e.g.*, Assistant Secretary of State Christopher Ford, "Strengthening Deterrence and Reducing Nuclear Risks: The Supplemental Low-Yield U.S. Submarine-Launched Nuclear Warhead," *Arms Control and International Security Papers*, vol. I, no. 4 (April 22, 2020), *available at* https://irp-cdn.multiscreensite.com/ce29b4c3/files/uploaded/ACIS%20Paper%204%20-%20W76-2%20Warhead.pdf.
49. *See, e.g.*, Assistant Secretary of State Christopher Ford, "Strengthening Deterrence and Reducing Nuclear Risks, Part II: The Sea-Launched Cruise Missile-Nuclear (SLCM-N)," *Arms Control and International Security Papers*, vol. I, no. 11 (July 23, 2020), *available at* https://irp-cdn.multiscreensite.com/ce29b4c3/files/uploaded/ACIS%20Paper%2011%20-%20SLCM-N.pdf.

weapons development, moreover, the United States must maintain and properly resource its *own* nuclear weapons infrastructure.[50]

Non-nuclear capabilities that also contribute to deterring aggression against U.S. allies must also be augmented so as to keep up with growing Sino-Russian threats. Now that the Intermediate-range Nuclear Forces (INF) Treaty has collapsed as a result of Russia's violation, for instance, conventionally-armed intermediate-range systems (whether land-, sea-, or air-launched) should be a particular priority. Such systems offer ways to degrade adversary "anti-access/area denial" (A2/AD) capabilities more easily and effectively than through exclusive reliance upon manned aircraft, thus preventing China and Russia from using their A2/AD and theater nuclear weapons to create envelopes of "denied space" projecting out from their borders over the territory of key U.S. allies, and under which aggression might be contemplated without fear of effective counter.[51] Effective deterrence relies upon being able to deny Beijing and Moscow such opportunities, and this will soon require more theater or regional conventional military capabilities than the United States and its allies presently possess.

Across the range of military capabilities, in fact – including next-generation tools facilitated by innovations in emerging technology fields such as human-machine teaming, as well as through steps to deter or counter Russian and Chinese militarization of outer

---

50. *See, e.g.*, Assistant Secretary of State Christopher Ford, "Deterrence and the U.S. Nuclear Weapons Infrastructure," *Arms Control and International Security Papers*, vol. I, no. 18 (September 9, 2020), *available at* https://irp-cdn.multiscreensite.com/ce29b4c3/files/uploaded/ACIS%20Paper%2018%20--%20Nuclear%20Infrastructure.pdf.

51. *See, e.g.*, Christopher Ford, "Offensive Nuclear Umbrellas and the Modern Challenge of Strategic Thinking," remarks at the U.S. House of Representative Visitors' Center to the Nuclear Security Working Group (February 10, 2016), *available at* https://www.newparadigmsforum.com/p2007.

space[52] and cyberspace[53] – it will be essential to ensure a robust, effective, and resilient U.S. and allied force posture. In light of the threats facing America's alliances, such capabilities are vital on their own terms, but we must not forget that as components of deterrence and of ally reassurance, they are also essential bulwarks against NPT withdrawal and nuclear weapons proliferation.

It will also be important for U.S. leaders to take very seriously their responsibilities in giving allies the security assurances they feel they need. It will not be enough, as may sometimes have happened in the past, for American officials to adopt the force posture and declaratory policies they want, and then lecture U.S. allies about how those choices should be reassuring to them. As threats grow in the security environment, power balances shift, and alliance solidarity comes to depend increasingly upon genuinely collective endeavor, we must also ensure that we truly listen to what our allies tell us – even if often only very privately – would be most reassuring to them, and we must make *their* desires and needs an important factor in our own procurement and planning, including in the nuclear arena.

If regional threats to U.S. allies grow especially severe, in fact, we must not be afraid to consider creative solutions about how to deter aggression and alleviate the nuclear weapons proliferation pressures that risk of aggression can create. When NATO Allies were confronted with grave Warsaw Pact threats in Central Europe

---

52. *See, e.g.*, Assistant Secretary of State Christopher Ford, "Arms Control in Outer Space: History and Prospects," *Arms Control and International Security Papers*, vol. I, no. 12 (July 24, 2020), *available at* https://irp-cdn.multiscreensite.com/ce29b4c3/files/uploaded/ACIS%20Paper%2012%20-%20Outer%20Space.pdf.
53. *See, e.g.*, Assistant Secretary of State Christopher Ford, "International Security in Cyberspace: New International Models for Reducing Risk," *Arms Control and International Security Papers*, vol. I, no. 20 (October 20, 2020), *available at* https://irp-cdn.multiscreensite.com/ce29b4c3/files/uploaded/ACIS%20Paper%2020%20--%20Cyberspace.pdf.

during the Cold War, for example, great effort was put into the development of NATO's "nuclear sharing" construct to help meet these needs.

Under NATO "nuclear sharing," the United States deployed nuclear weapons in Europe that were carefully kept under U.S. lock and key – and hence were not an NPT compliance problem under Article I,[54] as even Moscow acknowledged during the collaborative U.S.-Soviet negotiation of that provision[55] – but that were designated for potential release to and employment by multiple NATO allies in time of war, quite separately from the national nuclear forces maintained by the United Kingdom and France.[56] (NATO also established collaborative mechanisms for discussing nuclear weapons planning,[57] as well as for training NATO aircrews for potential nuclear delivery of U.S. weapons using dual-capable aircraft from national inventories certified for such use.[58]) Simultaneously confronting would-be invaders with the near-certainty of a NATO nuclear response in the event of invasion notwithstanding the emergence of strategic Soviet threats to U.S. cities, reassuring NATO allies such as West Germany that

---

54. NPT, *supra*, at Article I ("Each nuclear-weapon State Party to the Treaty undertakes not to transfer to any recipient whatsoever nuclear weapons or other nuclear explosive devices or control over such weapons or explosive devices directly, or indirectly; and not in any way to assist, encourage, or induce any non-nuclear-weapon State to manufacture or otherwise acquire nuclear weapons or other nuclear explosive devices, or control over such weapons or explosive devices.").
55. *See, e.g.*, Ford, "Challenges of Policymaking in Responsible Nuclear Weapons Stewardship," *supra*.
56. *See generally, e.g.*, "NATO's nuclear deterrence policy and forces," *NATO* (updated May 11, 2021), *available at* https://www.nato.int/cps/en/natohq/topics_50068.htm.
57. *See, e.g.*, NATO Archivist, *Nuclear Planning Group: 50th Anniversary of the First Meeting of the NPG* (2017), *available at* https://www.nato.int/nato_static_fl2014/assets/pdf/2020/3/pdf/200305-50Years_NPG.pdf.
58. *See, e.g.*, Rebecca Grant, "Nukes for NATO," *Air Force Magazine* (July 1, 2010), *available at* https://www.airforcemag.com/article/0710nato/.

they did not need their *own* nuclear weapons because U.S. devices would be made available in time of conflict, and reassuring the Soviets against the provocative development of additional individual *national* nuclear weapons programs in NATO states, these "nuclear sharing" innovations deftly supported deterrence, reassurance, nonproliferation, and crisis stability all at the same time and continue to support Alliance needs in the present day.[59]

Thankfully, there is so far still no need to contemplate an analogous arrangement of forward deployments and "dual-key" storage in the Indo-Pacific, nor for any major revision of such arrangements within NATO. We should remember this history, however, for it reminds us that careful planning and thoughtful diplomacy can sometimes discover ways forward that balance complex deterrence, reassurance, nonproliferation, and crisis stability needs. One hopes that such creativity is not needed in the future, but in worsening circumstances such concepts might still be better than the most likely alternative – specifically, weakened deterrence and increased proliferation pressure, potentially leading to conflict, to NPT withdrawal and indigenous nuclear weapons development by a U.S. ally, or to both of these things. From the perspective of deterring NPT withdrawal, strong U.S. alliances can help make *justified* departures from the Treaty framework unnecessary, thus allowing all states to focus more intently and directly upon deterring future North Korea-style problems.

V. Conclusion

The abovementioned twin-track agenda – for building like-minded coalitions of states committed to deterring destabilizing NPT withdrawal while shoring up U.S. alliances against Sino-Russian

---

59. *See, e.g.*, Ford, "Challenges of Policymaking in Responsible Nuclear Weapons Stewardship," *supra.*

threats in order to reduce the likelihood of withdrawal by one of America's allies or another state finding itself in the crosshairs of such potential aggression – is rather different than the more broadly internationalist approach advocated by Pierre Goldschmidt. Nevertheless, as long as the international community remains as divided as it has hitherto been on the withdrawal issue, as afflicted by proliferation-tolerant policies in Moscow and Beijing, and as paralyzed by multilateral timorousness and collective action problems, this twin-track approach may be the best one available.

The United States has partners who take nonproliferation seriously, and who seem to be deeply concerned about the possibility of further North Korea-style withdrawal from the NPT, particularly by Iran. These are states with whom we ought to be able to work in implementing such an agenda. We should not give up hope on more internationalist answers, but rather than simply waiting for the international community, as a whole, to do what it seems manifestly unwilling to do, we have before us an opportunity to use our own statesmanship to make at least progress in a dangerous world.

Toward an International Norm against Enrichment and Reprocessing Proliferation

Andrea Stricker

Abstract

The quinquennial review conference on the 1968 Nuclear Non-Proliferation Treaty (NPT), tentatively scheduled to take place in August 2022, provides the United States and its allies with an opportunity to make new commitments not to proliferate sensitive technologies that could contribute to the spread of nuclear weapons. The proliferation of advanced fuel cycle capabilities by such countries as Iran and North Korea, and the latter's development of atomic weapons, have created interminable obstacles to regional and international security. This paper recommends that at the upcoming NPT review conference, as a leader in creating and sustaining global norms, the United States should announce an official moratorium on providing new enrichment and reprocessing (E&R) capabilities – the key processes that produce nuclear explosive material – and seek the commitment of key partners to forgo further supply and novel E&R acquisition. In doing so, Washington can lay the groundwork for a reinvigorated, enduring policy to encourage a broader international interpretation that the NPT's intent is contrary to spreading E&R capabilities.

This paper suggests that U.S. and allied support for enhanced nonproliferation policy measures offers means to solidify a norm against new E&R acquisition, including offers of incentives for states not to acquire E&R technology, punitive sanctions, diplomatic pressure to halt, deter, or reverse possession, supply restrictions, and support for control regimes and safeguards. Where possible, the United States must reverse existing exceptions to E&R supply. With Washington's prolonged effort and sustained leadership, E&R acquisition and supply to additional states could rise to the level of a violation of an international norm.

NPT Articles

The dispute over which rights of technology acquisition the NPT grants its 191 parties is older than the treaty itself.[60] After several years of negotiations, the original drafters of the NPT ultimately left it ambiguous whether states should have access to E&R capabilities. In the years since, this controversy has remained front and center during inter-governmental struggles to control the supply of nuclear technology, and more recently, at NPT review conferences, where countries diverge on making new pledges to

---

60. Histories of NPT negotiations and compromises are available in: Henry Sokolski, Chapter 3, "What Does the History of the Nuclear Non-Proliferation Treaty Tell Us About Its Future?" in Sokolski, ed., Reviewing the Nuclear Nonproliferation Treaty (Washington, D.C.: Nonproliferation Policy Education Center, May 2010). (http://www.npolicy.org/books/ReviewingNPT/FullBook.pdf); Dean Rust, Chapter 2, "How We've Come to View the NPT: Three Pillars," in Sokolski, ed., *Nuclear Rules, Not Just Rights: The NPT Reexamined*, (Washington, D.C.: Nonproliferation Policy Education Center, February 2017. (http://npolicy.org/books/Nuclear_Rules_Not_Just_Rights/Ch2_Rust.pdf); Robert Zarate, "The NPT, IAEA Safeguards and Peaceful Nuclear Energy: An 'Inalienable Right,' But Precisely to What?" *Nonproliferation Policy Education Center,* Occasional Paper, May 8, 2007. (https://npolicy.org/the-npt-iaea-safeguards-and-peaceful-nuclear-energy-an-inalienable-right-but-precisely-to-what/)

strengthen such nonproliferation commitments.[61] Partly as a result, NPT states parties were unable to agree on an action plan to strengthen the treaty at the last review conference in 2015. Absent a consensus document, participants deemed the review conference a failure.[62]

Proponents of greater E&R restraint cite NPT Articles I and II as evidence that the treaty never intended to promote proliferation of sensitive technologies.[63] Indeed, not included in the treaty's text were proposals *requiring* countries with advanced nuclear technology to share with those that did not possess it, nor did it include mention of a *per se* right to such technology.[64]

Articles I and II, originally proposed by Ireland, became the treaty's key mechanisms to forestall further proliferation of atomic weapons while the nuclear weapon states (NWS) pursued eventual disarmament.[65] Article I clearly prohibits NWS from transferring nuclear weapons directly or indirectly and specifies that they may not "in any way to assist, encourage, or induce any non-nuclear-weapon State to manufacture or otherwise acquire nuclear weapons

---

61. See: David Albright and Andrea Stricker, "After the NPT Review Conference: Advancing the Non-Proliferation Pillar," *Institute for Science and International Security,* July 15, 2010. (https://isis-online.org/isis-reports/detail/after-the-2010-npt-review-conference-advancing-the-non-proliferation-pillar/23)
62. Louis Charbonneau, "Dispute Over Mideast Nuclear Arms Ban Torpedoes U.N. Conference," *Reuters,* May 22, 2015. (https://www.reuters.com/article/us-mideast-nuclear-un/dispute-over-mideast-nuclear-arms-ban-torpedoes-u-n-conference-idUSKBN0O72J220150523). The 2020 NPT review conference was postponed to 2022 due to the COVID-19 pandemic.
63. Treaty on the Non-Proliferation of Nuclear Weapons, 1968. (https://www.un.org/disarmament/wmd/nuclear/npt/text/)
64. Sokolski, Chapter 3, "What Does the History of the Nuclear Non-Proliferation Treaty Tell Us About Its Future?" in Sokolski, ed., *Reviewing the Nuclear Nonproliferation Treaty,* 2010; Zarate, "The NPT, IAEA Safeguards and Peaceful Nuclear Energy: An 'Inalienable Right,' But Precisely to What?" 2007.
65. Ibid.

or other nuclear explosive devices, or control over such weapons or explosive devices."

Article II restricts non-nuclear weapon state (NNWS) signatories from receiving transfer of nuclear weapons directly or indirectly. The NNWS also commit "not to manufacture or otherwise acquire nuclear weapons or other nuclear explosive devices; and not to seek or receive any assistance in the manufacture of nuclear weapons or other nuclear explosive devices."

Article III adds weight to an interpretation that states are free to acquire E&R capabilities under international safeguards, however, specifying:

> ...Safeguards required by this Article shall be implemented in a manner designed to comply with Article IV of this Treaty, and to avoid hampering the economic or technological development of the Parties or international co-operation in the field of peaceful nuclear activities, **including the international exchange of nuclear material and equipment for the processing, use or production of nuclear material for peaceful purposes**... (emphasis added)

Article IV either contravenes restraint arguments or supports them, depending on one's standpoint, by granting an "inalienable right of all the Parties to the Treaty to develop research, production and use of nuclear energy for peaceful purposes without discrimination," but adds that all must occur "*in conformity with Articles I and II*" (emphasis added). Article IV also stipulates, "All the Parties to the Treaty undertake to facilitate, and have the right to participate in, the fullest possible exchange of equipment, materials and scientific and technological information for the peaceful uses of nuclear energy."

Law experts such as Eldon Greenberg assess that "the rights and obligations established under Article IV can't be neatly disentangled from the prohibitions in Articles I and II of the Treaty. Indeed, Article IV contains express language linking the three provisions."[66] He continues, "In short, whatever the 'inalienable right[s]' enshrined in Article IV, they cannot be invoked to produce a result which, in the real world, entails unacceptable proliferation risks."[67] Greenberg concludes, "…Assistance or activities which are *ostensibly* peaceful in nature still are impermissible when *as a practical matter* they are likely to lead to the proliferation of nuclear weapons."[68] National security analyst Robert Zarate argues similarly, "The treaty can and should be interpreted in a pragmatic and sustainable way that rejects claims of the *per se* right of signatories to nuclear fuel-making…"[69]

The U.S. government has acknowledged that a straightforward reading of the NPT does not provide clarity for state policy, and therefore countries must make reasoned judgments about whether to supply proliferation-sensitive technology. In a 2005 arms control compliance report by the State Department and in corresponding Congressional testimony, for example, the department laid out several indicators that a state is violating Article II, or possibly undertaking the manufacture of nuclear weapons, and should not receive nuclear technology. A State Department official acknowledged that such deliberations naturally entail "judgments as to the purpose of a Party's nuclear

---

66. *Testimony of Eldon Greenberg on Assessing "Rights" Under the Nuclear Non-Proliferation Treaty,* before the Subcommittee on International Terrorism and Nonproliferation of the House Committee on International Relations, March 2, 2006.
67. Ibid.
68. Ibid.
69. Zarate, "The NPT, IAEA Safeguards and Peaceful Nuclear Energy: An 'Inalienable Right,' But Precisely to What?" 2007.

activities."[70] Greenberg also argues that certain conditions for supply must be met and "judgments…can and should be made on a case-by-case basis…"[71] Former U.S. Ambassador Christopher A. Ford views "safeguardability," or the ability to ensure against proliferation, as the supply approach that "reconciles the text of Article IV with the rest of the Treaty, with its negotiating record, and with longstanding international approaches to nuclear technology."[72] This has generally been the legal view taken by the State Department.

The easiest way to interpret the NPT, in the end, may be to imagine fictional outcomes at both extremes. Did the NPT aim to create 191 states with the ability to fabricate nuclear explosive material, leaving them very near to producing nuclear weapons? Further, did the NPT intend for most states to work on nuclear weaponization processes for atomic weapons, but stop short of assembling the weapons? Or would the preferable and intended outcome be that as few states as possible develop such capabilities and perform such work? States, by their own actions, have answered the question by tending toward the latter interpretation.

Moreover, states cannot rely on potentially flawed national intelligence assessments or judgments about whether to grant nuclear supply. They cannot predict what a state may do with E&R

---

70. U.S. Department of State, "Adherence to and Compliance with Arms Control, Nonproliferation, and Disarmament Agreements and Commitments," August 2005, explained by Christopher A. Ford in, "Iran, Nonproliferation, and the IAEA: A Legal History," *Hudson Institute,* November 2012, pp. 21-22. (https://www.hudson.org/content/researchattachments/attachment/1077/ford--iraniaeapaper1112.pdf)
71. Eldon V.C. Greenberg, "The NPT and Plutonium: Application of NPT Prohibitions to 'Civilian' Nuclear Equipment, Technology and Materials Associated with Reprocessing and Plutonium Use," *Nuclear Control Institute*, 1993.
72. Christopher Ford, "Nuclear Technology Rights and Wrongs: The NPT, Article IV, and Nonproliferation," *Nonproliferation Policy Education Center,* June 1, 2009. (http://npolicy.org/article.php?aid=351&rid=3)

technology at some future point, or how it might attempt to evade International Atomic Energy Agency (IAEA) safeguards.

Utilizing incentives and coercion, as appropriate, Washington can work to gain more adherents to a global understanding that the NPT does not advocate the spread of E&R capabilities and that a halt to further supply supports the aims of the treaty and most states' security objectives.

Status of E&R Proliferation

The interpretation that the NPT does not intend to promote E&R proliferation is supported by the behavior of states since its signing. In fact, the actions of most states signify greater advocacy for restraining the spread of E&R, rather than the reverse.

Presently, E&R capabilities globally are scarce even in countries that possess nuclear reactors or carry out nuclear research. This is both owing to practical deterrents against developing E&R technology, as well as normative ones relating to proliferation concerns. Each of these deterrents presents opportunities to grow a broader norm toward ending E&R proliferation.

About 14 countries enrich uranium, and several of them do so via regional consortiums.[73] States that enrich uranium principally pursue methods of centrifugation or gaseous diffusion. A handful of countries have developed other forms of enrichment, such as aerodynamic enrichment and laser separation. Enrichment is

73. World Nuclear Association, "Uranium Enrichment," Updated September 2020. (https://www.world-nuclear.org/information-library/nuclear-fuel-cycle/conversion-enrichment-and-fabrication/uranium-enrichment.aspx#:~:text=Most%20of%20the%20500%20commercial%20nuclear%20power%20reactors,process%20based%20on%20laser%20excitation%20is%20under%20development) The tally in this article should also include North Korea.

material-intensive and requires specialized expertise, including foreign assistance and knowledge.

France, Russia, and the Urenco consortium (Germany, the Netherlands, United Kingdom, and United States[74]) commercially enrich for domestic reactors and to produce and sell fuel cycle products. China currently enriches for its nuclear reactors but seeks to become a global fuel supplier. Argentina, Brazil, and Japan also have small enrichment programs for nuclear fuel, research, or naval propulsion. China, India, North Korea, and Pakistan operate enrichment programs that may actively contribute to nuclear weapons production.[75] Iran has a growing enrichment program that Tehran began to provide a source of weapons-grade uranium for nuclear weapons.[76] The program's stated civilian rationale remains in question.

The incidence of plutonium reprocessing is rarer. States that commercially reprocess plutonium include China, France, India, Japan, Russia, and the UK.[77] Nonproliferation expert Fred McGoldrick lists China, India, Israel, Pakistan, and North Korea as states that reprocess for nuclear weapons.[78] Along with China, the

---

74. The United States currently enriches uranium for domestic needs at a URENCO facility in New Mexico but is reviving its industry to support American advanced reactor sales to foreign countries.
75. Ruaridh Macdonald, "More Than We Need: Projected World Uranium Enrichment Capacity," *Nonproliferation Policy Education Center,* Occasional Paper 2106, October 2021. (https://npolicy.org/wp-content/uploads/2021/09/Estimates-of-Uranium.pdf)
76. David Albright, Olli Heinonen, and Andrea Stricker, "The Plan: Iran's Nuclear Archive Shows It Planned to Build Five Nuclear Weapons by Mid-2003," *Institute for Science and International Security,* November 20, 2018. (https://isis-online.org/isis-reports/detail/the-plan-irans-nuclear-archive-shows-it-originally-planned-to-build-five-nu/)
77. World Nuclear Association, "Processing of Used Nuclear Fuel," Updated December 2020. (https://world-nuclear.org/information-library/nuclear-fuel-cycle/fuel-recycling/processing-of-used-nuclear-fuel.aspx)
78. Fred McGoldrick, "Limiting Transfers of Enrichment and Reprocessing Technology: Issues, Constraints, Options" *Harvard University, Belfer Center for*

other four NPT-recognized NWS, the United States, France, Russia, and the UK, also retain the ability to reprocess for atomic weapons. China and the UK may be adding to their nuclear weapons arsenals.[79]

States have tried to reduce the justification for other countries to develop the full fuel cycle by providing affordable and predictable access to nuclear reactor fuel. For example, in 2019, the IAEA inaugurated the world's first international nuclear fuel bank in Kazakhstan to allow countries to purchase low enriched uranium fuel if it is unavailable from other sources.[80] Such initiatives can assist norm development against wider E&R propagation, insofar as they remove justification for states to make their own fuel.

Incentives and Disincentives

Parties to the NPT that view E&R proliferation as counter to the treaty's intent have addressed the treaty's silence through other means. This underscores that a collective of states already hold the interpretation that the propagation of advanced fuel cycle technologies is inconsistent with the NPT's aim to prevent the spread of nuclear weapons and encourage their dismantlement.

Science and International Affairs, Project on Managing the Atom, May 2011, p. 10.
(https://scholar.harvard.edu/files/matthew_bunn/files/limiting_transfers_of_enrichment_reprocessing_technology.pdf)

79. Ryan Browne, "China is Aiming to Double the Size of its Nuclear Arsenal, Pentagon Report Says," *CNN,* September 1, 2020.
(https://www.cnn.com/2020/09/01/politics/china-military-power-report/index.html); "Integrated Review: UK to Lift Cap on Nuclear Stockpile," *BBC News,* March 16, 2021. (https://www.bbc.com/news/uk-56413920)

80. Catherine Putz, "Low Enriched Uranium Fuel Bank Officially Operational in Kazakhstan," *The Diplomat,* October 29, 2019.
(https://thediplomat.com/2019/10/low-enriched-uranium-bank-officially-operational-in-kazakhstan/)

One incentive for states to forgo E&R development is the prestige and elevation in relations that comes with concluding a bilateral nuclear cooperation agreement with the United States which formalizes a commitment not to enrich or reprocess – known colloquially as the "gold standard" of nonproliferation. The United States has utilized this approach to some success, where thus far two states – the United Arab Emirates (UAE) and Taiwan – have pledged not to develop E&R. Such agreements are concluded under section 123 of the U.S. Atomic Energy Act (AEA). The UAE's commitment runs through 2039 and Taiwan's is indefinite. Vietnam also committed in the preamble of its 123 agreement not to enrich or reprocess, and its accord runs through at least 2044.[81] A similar agreement with Mexico, which has not yet entered into force, will reportedly contain a similar commitment and the accord will endure for 30 years.[82] The U.S./UAE accord has contributed most markedly to deepened political and security relations, including paving the way for significant U.S. military sales to the UAE. The UAE case offers a template for how prestige and security via close relations with America can more easily flow from E&R restraint.

States' desire to promote anti-E&R regional commitments and bolster their nonproliferation *bona fides* presents opportunities for Washington and its partners to further develop a norm against E&R acquisition. The UAE's stance offers an important counterpoint to additional Middle Eastern states seeking enrichment and reprocessing to gain parity with Iran. Taiwan's

---

81. Paul K. Kerr and Mary Beth D. Nikitin, "Nuclear Cooperation with Other Countries: A Primer," *Congressional Research Service,* Updated September 21, 2020. (https://fas.org/sgp/crs/nuke/RS22937.pdf)
82. President Donald J. Trump, "Message to the Congress Transmitting a Proposed Agreement for Cooperation Between the Government of the United States of America and the Government of the United Mexican States Concerning Peaceful Uses of Nuclear Energy," May 8, 2018. (https://www.presidency.ucsb.edu/documents/message-the-congress-transmitting-proposed-agreement-for-cooperation-between-the-4)

decision enhances its stature as a responsible nonproliferation actor vis-à-vis China. While the U.S. Intelligence Community assesses that China is producing additional nuclear weapons, Taipei appears a reformed and restraint-favoring actor. Its anti-E&R commitment followed a clandestine nuclear weapons program and covert plutonium reprocessing effort, which U.S. pressure succeeded in closing down in the late 1980s.[83] For the UAE, committing not to acquire E&R also assisted Abu Dhabi in rehabilitating its nonproliferation image after functioning as a long-time hub of proliferant states' illicit procurement and financial activity.

The U.S. AEA also empowers Washington to discourage the spread of E&R, requiring countries with which the United States shares nuclear technology to have Congressional approval for enrichment or reprocessing using U.S.-origin nuclear fuel and technologies. Washington is legally required to terminate nuclear cooperation if a state violates its nonproliferation commitments.[84] Other U.S. laws support the nonproliferation of sensitive technologies, such as the Nuclear Non-Proliferation Act of 1978 and the Nuclear Proliferation Prevention Act of 1994, the latter of which "imposes economic and military assistance sanctions on countries supplying and receiving enrichment and reprocessing equipment, materials and technology."[85]

International disincentives for E&R acquisition are frequently more high profile than incentives. Iran, Iraq, Libya, and North Korea all sought nuclear weapons via secret enrichment or reprocessing programs, and the UN Security Council passed

---

83. David Albright and Andrea Stricker, *Taiwan's Former Nuclear Weapons Program: Nuclear Weapons On-Demand,* (Washington, D.C.: Institute for Science and International Security, 2018). (https://isis-online.org/books/detail/taiwans-former-nuclear-weapons-program-nuclear-weapons-on-demand/15)
84. Kerr and Nikitin, "Nuclear Cooperation with Other Countries: A Primer."
85. *Testimony of Eldon Greenberg,* March 2, 2006.

numerous resolutions demanding the suspension or dismantlement of those programs and imposed sanctions. The UN Security Council, therefore, plays a significant role in interpreting the NPT's silence on E&R by issuing explanatory resolutions about permissible activities, especially when states' activities appear geared toward weapons development.[86]

States have also employed unilateral and secondary sanctions, as well as trade restrictions, to deter states from moving forward with E&R programs. Israel has deterred E&R acquisition through kinetic action, destroying reactors in Iraq and Syria that would have produced plutonium for nuclear weapons, and sabotaging Iran's enrichment program.

Natural deterrents to countries seeking home-grown or imported E&R include the fact that states can purchase reactor fuel readily and cheaply from established suppliers rather than fabricate their own. Nuclear scientist Ruaridh Macdonald assesses that the projected increase in demand for enriched uranium fuel for power reactors, for example, is likely to be met by existing suppliers due to states' "significant overinvestment in enrichment capacity during the 90s and first decade of this century…"[87]

Should states seek to proliferate E&R capabilities covertly, they face obstacles in purchasing such facilities from a state supplier, and raw materials and components either from a state or on the black market. They must also acquire outside expertise and spend years building a domestic technological base through time-intensive training acquired from abroad. Many advanced nuclear

86. Orde Kittrie, "Averting Catastrophe: Why the Nuclear Nonproliferation Treaty is Losing its Deterrence Capacity and How to Restore It," *Michigan Journal of International Law,* Vol. 28, 2007, pp. 337-430. (https://papers.ssrn.com/sol3/papers.cfm?abstract_id=996953)
87. Macdonald, "More Than We Need: Projected World Uranium Enrichment Capacity," p. 4.

technology states closely monitor and seek to prevent the leakage of such know-how.

States' desire not to raise proliferation suspicions of neighbors ideally disincentivizes them from acquiring E&R, but only if other security concerns are not paramount. States realize that developing advanced fuel cycle capabilities positions them closer to nuclear weapons and leaves their adversaries less secure and more prone to weighing their options. Since the exposure of Iran's covert enrichment and reprocessing programs in 2003, Tehran has triggered concerns of neighbors – concerns compounded by the fact that world powers failed to negotiate the programs' permanent termination. At least two countries, Saudi Arabia and Turkey, have stated publicly that they might be driven to match Tehran, and Riyadh may be developing uranium production facilities.[88] Jordan and Egypt will also not rule out E&R development.

The UAE, by taking the opposite approach and openly committing not to seek E&R, even as it acquires reactors for nuclear energy production, assures its neighbors that it will not participate in a dangerous regional fuel cycle race which could quickly turn into a nuclear arms race.[89] Although Egypt has considered nuclear weapons and refuses to undertake enhanced nonproliferation

---

88. David E. Sanger, "Saudi Arabia Promises to Match Iran in Nuclear Capability," *The New York Times,* May 13, 2015. (https://www.nytimes.com/2015/05/14/world/middleeast/saudi-arabia-promises-to-match-iran-in-nuclear-capability.html); "Erdogan Says its Unacceptable that Turkey Can't Have Nuclear Weapons," *Reuters,* September 4, 2019. (https://www.reuters.com/article/us-turkey-nuclear-erdogan-idUSKCN1VP2QN; Andrea Stricker and Behnam Ben Taleblu, "Secret Sites in the Desert: the Dangers of Saudi Arabia's Nuclear Hedging," *The National Interest,* September 4, 2020. (https://nationalinterest.org/feature/secret-sites-desert-dangers-saudi-arabia%E2%80%99s-nuclear-hedging-168267)
89. See, for example: "Statement of UAE Ambassador Yousef Al Otaiba on Barakah Plant Unit 1 Commercial Operations," April 6, 2021. (https://www.uae-embassy.org/news-media/ambassador-yousef-al-otaiba-statement-barakah-plant-unit-1-commercial-operations)

commitments, it has also spearheaded the creation of a Middle East zone free of nuclear weapons. Both countries provide precedents for Saudi Arabia and Turkey to emulate, although it remains to be seen if they will follow suit. If Tehran's enrichment program continues, along with unresolved nuclear weapons dimensions in Iran, Riyadh and Ankara may address their insecurity by seeking E&R to position themselves closer to nuclear weapons.

The prevalence of disincentives for E&R development and incentives for restraint nevertheless underscores a preeminent international view that the NPT does not aim to spread advanced fuel cycle technology and most states do not interpret it to do so.

Supply Constraints and Enhanced Safeguards

States have worked over decades to create technology control groups and regimes around the preponderant consensus that it is in the global interest to hinder additional states from acquiring E&R technologies. States have also sought strengthened IAEA safeguards to detect potential NPT violations.

Following the NPT's entry into force in 1970, for example, between 1971 and 1974, a group of 15 nuclear exporting states – not all NPT members at the time – established the Zangger Committee to reach common understandings on:

> (a) the definition of what constituted 'equipment or material especially designed or prepared for the processing, use or production of special fissionable material' (as it was not defined anywhere in the Treaty);
>
> (b) the conditions and procedures that would govern exports of such equipment or material in order to meet

the obligations of Article III, paragraph 2 of the NPT, on a basis of fair commercial competition.[90]

The group agreed on conditions for exporting nuclear equipment and materials, which evolved into guidelines known as a "trigger list." The trigger list includes specific direct- and dual-use materials and equipment that activate nations' supply controls. The Zangger Committee currently has a membership of 39 NPT member states, which meet to discuss changes and additions to the trigger list.

The Zangger Committee trigger list was further formalized by a group of seven states that met from 1975 to 1978, later called the Nuclear Suppliers Group (NSG). NSG member states agreed to embrace the Zangger trigger list as the basis for controlling direct-use nuclear supply, including nuclear facilities, materials, technology, and equipment.

In 1978, IAEA member states adopted the NSG's Part I list, requiring IAEA safeguards for a state to receive such commodities.[91] The current NSG guidelines include specific conditions for transferring enrichment and reprocessing capabilities, for example, NPT membership, good nonproliferation standing with the IAEA and UN Security Council, an IAEA Additional Protocol or other appropriate safeguards in force, and legally binding commitments that a state will not enrich uranium to greater than 20 percent purity. NSG members also agreed to encourage other options over E&R, such as a state's participation in a multinational arrangement to help it avoid needing direct

---

90. Zangger Committee, "History."
http://www.zanggercommittee.org/history.html
91. Nuclear Suppliers Group, "About the NSG."
https://www.nuclearsuppliersgroup.org/en/about-nsg

access to the technology.[92] In 1992, the NSG agreed to additional guidelines to control transfers of dual-use materials, technology, and equipment, formalized in the NSG Part II list, which the IAEA also adopted. The NSG lists seek "to ensure that nuclear trade for peaceful purposes does not contribute to the proliferation of nuclear weapons or other nuclear explosive devices, and that international trade and cooperation in the nuclear field is not hindered unjustly in the process."[93]

The NSG has grown to 48 members which meet regularly to discuss updates to the control lists and conditions for supply. NSG membership itself requires NPT membership or membership in a nuclear weapon-free-zone treaty and application of full-scope IAEA safeguards. Since the group operates by consensus, strengthening measures have represented hard-won victories. Even responsible states regarding nonproliferation often resist supply constraints affecting future E&R commercial interests. Other states seek to preserve their option to achieve parity with neighboring countries.[94]

In another norm-reinforcing measure, in 2004, the UN Security Council passed Resolution 1540 to prevent the transfer of dual-use commodities to non-state actors that could enable them to build weapons of mass destruction. Resolution 1540 also required states to establish national controls over sensitive technologies, equipment, and know-how.[95] As of 2021, however, according to a

---

92. "Guidelines for Nuclear Transfers: NSG Part I Guidelines," INFCIRC/254/Rev. 14/Part 1. https://www.iaea.org/sites/default/files/publications/documents/infcircs/1978/infcirc254r14p1.pdf
93. Nuclear Suppliers Group, "Guidelines." https://www.nuclearsuppliersgroup.org/en/guidelines
94. McGoldrick, "Limiting Transfers of Enrichment and Reprocessing Technology," pp. 19-24.
95. UN Security Council Resolution 1540, 2004, (https://www.un.org/ga/search/view_doc.asp?symbol=S/RES/1540%20(2004))

comprehensive analysis by the Institute for Science and International Security, only 76 countries have adequate export control legislation in place and implementation remains deficient.[96]

States' efforts to strengthen IAEA safeguards over time also indicate a majority view toward concern about advanced fuel cycle activities. During the 1980s and 1990s, Iraq, North Korea, Libya, and Taiwan covertly established or furthered nuclear weapons programs despite having comprehensive safeguards agreements (CSAs) with the IAEA, as required by the NPT. A quasi-state-backed nuclear technology proliferation network operating out of Pakistan, the A.Q. Khan network, had supplied sensitive nuclear technology to several of these countries, including Iraq, Libya, and North Korea (which then shared it with Syria). The Khan network also supplied South Africa and Iran, and perhaps others.[97]

In response, from the early- to mid-1990s, the IAEA Board of Governors formally acknowledged that the IAEA has not only an obligation to ascertain the "correctness" of states' CSA declarations, but also their "completeness."[98] In 1997, the Board authorized states to conclude a voluntary, supplementary agreement to the CSA, known as the Additional Protocol or AP.[99] The AP enables "the IAEA inspectorate to provide assurance about

---

96. David Albright, Sarah Burkhard, Spencer Faragasso, and Linda Keenan, *The Peddling Peril Index 2021/2022: Ranking Strategic Trade Controls* (Washington, D.C.: Institute for Science and International Security Press, 2021).
97. David Albright, *Peddling Peril: How the Secret Nuclear Trade Arms America's Enemies* (New York: Free Press, 2010).
98. David Fisher, *The History of the International Atomic Energy Agency: The First Forty Years* (Vienna: IAEA, 1997). (https://www-pub.iaea.org/MTCD/publications/PDF/Pub1032_web.pdf); David Albright, Olli Heinonen, and Orde Kittrie, "Understanding the IAEA's Mandate in Iran: Avoiding Misinterpretations," *Institute for Science and International Security,* November 27, 2012. (https://isis-online.org/uploads/isis-reports/documents/Misinterpreting_the_IAEA_27Nov2012.pdf)
99. IAEA, *Model Additional Protocol to the Agreement(s) Between State(s) and the International Atomic Energy Agency,* INFCIRC/540 (Corrected), September 1997. (https://www.iaea.org/sites/default/files/infcirc540c.pdf)

both declared and possible undeclared activities. Under the Protocol, the IAEA is granted expanded rights of access to information and sites" provided under the CSA.[100] To date, 137 of the IAEA's 172 member states and Euratom have an AP in force, reflecting a majority in favor of strengthened safeguards.[101]

The task for Washington and likeminded partners is to further enhance and propagate this common view that the NPT does not promote proliferation of E&R and translate it into a norm against further supply.

Building a Norm

How does a policy goal become an international norm? According to international law experts, norms "express what is esteemed and valued and are means to realize valuable end states." The key is that "norms must be shared with other actors, and partly enforced by their sanctions."[102] A primary goal of the United States and its partners, then, as the "norm entrepreneurs" "is to define and/or establish specific standards of behavior with the expectation that other actors within the community follow them. They might also try to replace an old standard of behavior with a new one."[103] With

---

100. IAEA, "IAEA Safeguards Overview." (https://www.iaea.org/publications/factsheets/iaea-safeguards-overview)
101. IAEA, "Additional Protocol." (https://www.iaea.org/topics/additional-protocol)
102. Tore Nyhamar, "How Do Norms Work? A Theoretical and Empirical Analysis of African International Relations," *International Journal of Peace Studies,* Vol. 5, No. 2, Autumn/Winter 2000, (https://www.gmu.edu/programs/icar/ijps/vol5_2/nyhamar.htm)
103. Martha Finnemore and Kathryn Sikkink, "International Norm Dynamics and Political Change," *International Organization*, Vol. 52, No. 4, Autumn 1998, pp. 887-917. (https://home.gwu.edu/~finnemor/articles/1998_norms_io.pdf)

time, "Treaties can [play] a significant role in the development of customary international law," or what follows norm creation.[104]

Academics Michal Smetana and Joseph O'Mahoney argue that a well-developed norm against the spread of nuclear weapons underpins the nonproliferation regime and the NPT, and as such, the nonproliferation regime amounts to an "anti-fragile system" that has successfully resisted contestation and assaults to its integrity. For example, despite several close calls and states developing nuclear latency, or proximity to nuclear weapons, the NPT has lost only one signatory – North Korea – to proliferation.[105]

Smetana and O'Mahoney also assert that there is no "quasi-automatic process in which a norm violation leads to the erosion or weakening of the norm in question."[106] They cite Professor Nassim Nicholas Taleb's assertion that an antifragile system can "benefit from shocks [and] grow when exposed to volatility, randomness, disorders, and stressors."[107] Contestations of nonproliferation norms, then, provide chances for states to reaffirm norms and clarify boundaries. However, Smetana and O'Mahoney acknowledge that "chronic and frequent norm violations would

---

104. Rudolph Bernhardt, ed. *Encyclopedia of Public International Law* (Netherlands: North-Holland Elsevier, 2000), pp. 951-952.
105. North Korea left the NPT in 2003. Israel, India, and Pakistan did not sign the NPT and possess nuclear weapons. South Africa also did not join the NPT and developed and then dismantled nuclear weapons, signing the NPT in 1991 as a NNWS.
106. Michal Smetana and Joseph O'Mahoney, "NPT as an Antifragile System: How Contestation Improves the Nonproliferation Regime," *Contemporary Security Policy,* September 16, 2021. (https://doi.org/10.1080/13523260.2021.1978761)
107. Nassim Nicholas Taleb, *Antifragile: Things that Gain from Disorder* (New York: Random House, 2014).

certainly threaten the regime's effective functioning and lead to maladaptive outcomes..."[108]

While resilience may describe the nonproliferation regime and the NPT, how well developed, by comparison, is E&R restraint as a norm? Within a norm "life cycle," according to international relations scholars Martha Finnemore and Kathryn Sikkink, the first stage, "norm emergence," is already underway. States have recognized for decades that E&R proliferation is a threat to international security and have taken steps to prevent and deter it. The second stage is also in progress, known as "a dynamic of imitation...as the norm leaders attempt to socialize other states to become norm followers." The NSG, for example, serves as one potential platform for socialization, but many of its members refuse to foreswear the option to acquire E&R. However, a norm against further E&R proliferation has not reached the third stage, "norm internalization," where norms "acquire a taken-for-granted quality and are no longer a matter of broad public debate."[109]

By implication, a norm against the spread of E&R capabilities can only reach the third stage, or fruition, after consistent action. This must involve clarity by states that they will not sell or assist other states in acquiring the technology, and commitments by states without E&R that they will not seek it. Importantly, states must refrain from creating new exemptions for E&R proliferation or they will continue to undermine norm development. Norm entrepreneurs must regularly coordinate actions to maintain progress on establishing the norm.

For an anti-E&R norm to emerge, states must take strengthening actions across many areas of incentives, disincentives, supply constraints, and safeguards. The United States and its partners

---

108. Smetana and O'Mahoney, "NPT as an Antifragile System."
109. Ibid.

must work through the UN, NSG, Resolution 1540 committee, and other fora, as well as unilaterally, to establish a global halt to further E&R supply. They must counter violations through pressured diplomacy and sanctions. States must seek to prevent the spread of new E&R capabilities, regardless of whether the technology seeker is a friend or an adversary.

Undermining Norm Creation

The NSG, while playing an indispensable role in strengthening nuclear export controls, has not been able to agree on halting all further supply of E&R, raising doubt about the group's ability to perpetuate an anti-E&R norm to fruition. Since the NSG operates by consensus, it has proven a contentious venue for building a norm against further E&R supply.[110] Not all members possess E&R and seek to retain the option to acquire it based on their NPT "right." Several have also resisted introducing stronger "subjective" supply criteria to NSG guidelines. In 1994, the NSG did adopt a "nonproliferation principle," requiring suppliers to "not contribute to the proliferation of nuclear weapons."[111]

From 2004 to 2011, the NSG endeavored to adopt stronger criteria for E&R supply. In 2004, following the unraveling of the Khan network, the George W. Bush administration supported a norm against the further spread of E&R by proposing that NSG members agree to formally halt all new supply to countries that did not already possess the technology while providing readily available

110. See in particular: Mark Hibbs, "The Future of the Nuclear Suppliers Group," *Carnegie Endowment for International Peace,* 2011. (https://carnegieendowment.org/files/future_nsg.pdf)
111. Nuclear Suppliers Group, "About the NSG." (https://www.nuclearsuppliersgroup.org/en/about-nsg)

nuclear fuel to meet their energy needs.[112] In 2011, the Obama administration settled for member states' consensus to strengthen the NSG's "objective" criteria-based approach for supply. Recipients would require a CSA with the IAEA, an IAEA Additional Protocol or "pending this…appropriate safeguards agreements," [113] good standing with their IAEA safeguards obligations, compliance with UNSCR 1540, a written agreement of non-diversion, adequate accounting and control over nuclear materials, and appropriate physical protection and safety standards. Suppliers should also seek legally binding agreements that recipients will not modify or operate an enrichment plant to produce greater than 20 percent enriched uranium and to design facilities in a way that precludes this.[114]

Despite U.S. leadership, even NSG countries with strong nonproliferation records did not want their option to acquire E&R foreclosed.[115] Several refused to adopt Washington's proposed "black-box" criteria to prevent replication of E&R technologies. Ultimately, the United States agreed to black-box criteria applying only to E&R technologies or facilities possessed by states prior to

---

112. "Remarks by the President on Weapons of Mass Destruction Proliferation," February 11, 2004. (https://georgewbush-whitehouse.archives.gov/news/releases/2004/02/20040211-4.html)
113. Argentina and Brazil refuse to sign an Additional Protocol, viewing as sufficient their participation in a mutual inspection agreement, of which the IAEA is also a member. NSG consensus required adding language on "appropriate safeguards agreements." See: Arms Control Association, "NSG Revises Rules on Sensitive Exports," June 27, 2011, Updated July 5, 2011. (https://www.armscontrol.org/arms-control-today/2011-07/nsg-revises-rules-sensitive-exports)
114. Nuclear Suppliers Group, "NSG Part I Guidelines," INFCIRC/254/Rev.14/Part 1a. (https://www.iaea.org/sites/default/files/publications/documents/infcircs/1978/infcirc254r14p1.pdf)
115. For a concise summary of various NSG member positions, see: McGoldrick, "Limiting Transfers of Enrichment and Reprocessing Technology," pp. 13-18. Argentina, Brazil, Canada, the Netherlands, Spain, South Korea, Switzerland, South Africa, and Turkey took strong positions.

December 31, 2008, but the criteria are somewhat weak. They contain language such as "avoid, as far as practicable" and "seek from recipients" legal agreements preventing replication. New criteria also specify that countries which acquire *new* E&R capabilities should endeavor to ensure they are not replicable once commercially available. States rejected a provision that countries which had already committed not to acquire E&R could not receive the technology. Several states also sought to avoid introducing "subjective" criteria, such as suppliers assessing the credibility of a potential recipient's given rationale for seeking E&R or the potential for supply to fuel regional instability. Instead, the updated guidelines specify that countries should generally "tak[e] into account at their discretion, any relevant factors as may be applicable."[116]

NSG members also agreed to language indicating, "All States that meet the criteria...are eligible for transfers of enrichment facilities, equipment and technology." A U.S. official told the Arms Control Association that Washington, however, did not view eligibility as equating to a "right" to E&R supply.[117] Other NSG members clearly disagree with this interpretation.

Thus, while the NSG does not actively undermine anti-E&R norm building, some of the body's 48 members – including close U.S. allies – are clearly not prepared to support it. This conflict underscores the continued tension created by the NPT's reference to an "inalienable right" to develop nuclear energy for peaceful purposes, and the treaty's lack of clarity on which technologies are permissible. As such, the United States may need to work outside the NSG platform, but potentially with some NSG partners, to cultivate an anti-E&R norm.

---

116. Nuclear Suppliers Group, "NSG Part I Guidelines."
117. Arms Control Association, "NSG Revises Rules on Sensitive Exports."

The United States has also undermined the potential for an anti-E&R norm by permitting key exceptions to supply on at least two recent occasions. Where possible, as a norm entrepreneur, it should consider attempting to reverse those exceptions.

In 2007, the United States contravened its long-standing nonproliferation policies by permitting India – a state outside the NPT that has nuclear weapons – to sign a 123 agreement allowing New Delhi to import U.S.-origin natural and low enriched uranium, carry out reprocessing using U.S.-origin plutonium, and even construct reprocessing plants. The plants require IAEA safeguards and are to be used strictly for nuclear energy applications.[118] Yet, since India enriches and reprocesses for nuclear weapons, in effect, Washington allowed New Delhi to devote its existing resources entirely to nuclear weapons production, if it so chose. Further, in 2008, following a U.S.-led campaign, the NSG granted India an exemption to receive E&R technology without being a member of the NPT or a nuclear weapon-free-zone and applying IAEA safeguards. The United States supported India receiving a waiver for NSG membership, but China blocked its entry.

The second major exception involved Iran. In 2015, the P5+1, or the United States, France, Germany, the UK, Russia, and China, concluded a deal with Tehran that permitted it to maintain its formerly covert enrichment program. Iran had acquired the program illicitly by importing needed equipment and expertise, including drawing on substantial assistance from the Khan network.[119] Between 2006 and 2010, the Security Council passed

---

118. *Agreement for Cooperation between the Government of the United States of America and the Government of India Concerning Peaceful Uses of Nuclear Energy (123 Agreement),* Published August 3, 2007. (https://2001-2009.state.gov/r/pa/prs/ps/2007/aug/90050.htm)
119. Albright, *Peddling Peril: How the Secret Nuclear Trade Arms America's Enemies.*

successive resolutions demanding Iran suspend its enrichment and reprocessing programs.

The Iran nuclear deal, or Joint Comprehensive Plan of Action (JCPOA), and associated UN Security Council resolution 2231 superseded the prior resolutions and specified only temporary restrictions on Iran's enrichment program. After 15 years, Iran may grow its enrichment program to industrial levels and faces no restrictions on reprocessing.[120] It may also amass weapons-grade uranium. Other countries will be able to conduct nuclear trade with Iran after ten years, or in 2025. Washington left the JCPOA in 2018 under President Donald Trump, who stated that Iran must abandon enrichment to receive sanctions relief. Iran then began violating the accord in May 2019 and left it in January 2020.[121] As of January 2022, the Biden administration and its five counterparts are seeking to revive the agreement and obtain Iran's commitment to most of the deal's original terms.[122]

Japan represents another long-standing E&R exception that Washington might seek to address. Tokyo has a pre-existing enrichment program but cannot use U.S.-origin material and equipment for enrichment over 20 percent purity unless it receives approval. It may also reprocess plutonium transferred under the accord if Washington approves it, which U.S. presidents have

---

120. *Joint Comprehensive Plan of Action,* Vienna, July 14, 2015. (https://2009-2017.state.gov/e/eb/tfs/spi/iran/jcpoa/index.htm)
121. Michael R. Gordon, "U.S. Lays Out Demands for New Iran Deal," *The Wall Street Journal,* May 21, 2018. (https://www.wsj.com/articles/mike-pompeo-lays-out-next-steps-on-iran-1526909126); Max Burman and The Associated Press, "Iran Pulling Out of Nuclear Deal Commitment After U.S. Strike that Killed Soleimani," *NBC News,* January 5, 2020. (https://www.nbcnews.com/news/world/iran-pulling-out-nuclear-deal-following-u-s-strike-killed-n1110636)
122. Pariza Hafezi and John Irish, "Iran, U.S. Lock Horns Over Sanctions Relief, Nuclear Curbs in Vienna Nuclear Talks," *Reuters,* January 12, 2022. (https://www.reuters.com/world/middle-east/iran-us-lock-horns-over-sanctions-relief-nuclear-curbs-vienna-talks-2022-01-12/)

regularly done.[123] By comparison, the United States has not granted E&R rights in a U.S. 123 agreement to South Korea, despite having comparatively close relations with the country.[124] Seoul, which seeks pyroprocessing for spent fuel management, certainly observes the United States effectively granting India, a state with nuclear weapons outside the NPT, expanded E&R programs. It also witnesses America concluding a deal with Iran for continued enrichment. All of this has occurred despite the restraint South Korea has shown toward North Korea's possession of E&R technology and nuclear weapons. Instead of rectifying past E&R exemptions, recent U.S. efforts to bring pyroprocessing to fruition with South Korea signify that Washington may be considering yet another exception.[125]

Such further exceptions to E&R supply undermine norm creation. To rectify these missteps, the United States could seek to renegotiate or cancel the U.S./India deal, but would have to do so in part legislatively. At a minimum, Washington should refrain from further support for India's NSG membership. It should also insist on a new accord with Iran that restores the principle of non-enrichment and reprocessing as a condition for unilateral and multilateral sanctions relief, as well as hold Tehran accountable to full transparency with the IAEA over the past and potentially ongoing military dimensions to its nuclear program. The United States would need to gain consensus of the other P5+1 countries to abandon the JCPOA and UNSCR 2231 and restore unity of

---

123. *Agreement for Cooperation Between the Government of the United States of America and the Government of Japan Concerning Peaceful Uses of Nuclear Energy,* July 17, 2018. (https://fissilematerials.org/library/jp123.pdf)
124. Ibid, p. 7. Caveats are that the U.S. permitted Japan to send spent reactor fuel to Europe for reprocessing and has authorized a limited form of reprocessing cooperation which it does not view as sensitive. See: McGoldrick, "Limiting Transfers of Enrichment and Reprocessing Technology," pp. 9-10.
125. Hae-Sung Lee, "US, Korea Approve Nuclear Fuel Recycle Technology," *The Korea Economic Daily,* September 2, 2021. (https://www.kedglobal.com/newsView/ked202109020004)

response to Iran's provocative nuclear advances. These goals represent no small diplomatic feats.

If the United States rejoins the Iran nuclear deal and agrees to lift sanctions in return for expiring restrictions on Iranian enrichment, by contrast, it must anticipate and counter E&R acquisition by Iran's neighbors. South Korea may also seek E&R on its own or via alternative state suppliers to get closer to a nuclear weapons option if Washington reaches a limited nuclear rollback agreement with North Korea and abandons its long-standing principle of complete and verified denuclearization of the peninsula. While the United States is attempting to balance with Seoul prior exceptions made for Tokyo, pyroprocessing technology could also offer South Korea a first step toward gaining access to reprocessing know-how.

While reversing E&R exceptions and terminating progress toward any new exceptions may be arduous and painful, the costs of not doing so may be far worse in terms of proliferation in the Middle East and Northeast Asia.

Announcing a Moratorium

Ahead of the 2022 NPT review conference, the United States should formally announce a moratorium on new authorizations of enrichment and reprocessing in U.S. nuclear cooperation agreements. It should seek partners in promoting the interpretation that the NPT does not encourage the development of advanced fuel cycle capabilities, enlisting states that will commit publicly to a statement forgoing new E&R acquisition and supply for the foreseeable future. A statement could include these states' commitment to not reevaluate their decision until a far-off date, such as 2040, in order to provide space for norm development.

Washington should offer lucrative incentives for partners to join this coalition. Even a limited number of partners agreeing to a new commitment would support norm-building efforts.

Importantly, this policy will need to endure across successive U.S. administrations, so it will be imperative to obtain support from career bureaucrats.

Washington must make the case to states that currently possess E&R technology, such as European enrichment and reprocessing states, that they too should not provide new E&R supply. With an expected uptick in nuclear fuel demand, it may also be in these states' economic interest to maintain market control. The United States may be able to successfully make this case to two P5 members, France and the UK, but would almost certainly not obtain Russia and China's endorsement.

As such, the United States must plan to counter new E&R supply by these revisionist states, particularly in light of worsening relations and more willingness by Moscow and Beijing to flout international norms. Washington should consider sanctioning official state entities that assist partners in developing E&R, such as if China's alleged front-end fuel cycle cooperation with Saudi Arabia deepens to entail E&R support or if Russia assists Iranian nuclear proliferation.[126] The United States must maintain rigorous diplomatic and economic overtures to allies and partners, including providing security assurances, to dissuade them from purchasing Russian or Chinese E&R technologies.

---

126. Russian scientists assisted Iran's nuclear weapons program in the past. See: David Albright and Robert Avagyan, "Revisiting Danilenko and the Explosive Chamber at Parchin: A Review Based on Open Sources," *Institute for Science and International Security,* September 17, 2012. (https://isis-online.org/isis-reports/detail/revisiting-danilenko-and-the-explosive-chamber-at-parchin-a-review-based-on/)

The past difficulty of achieving agreement within the NSG on a set of strong anti-E&R criteria indicates that the United States should work outside the group to cultivate a subset of nations that commit not to provide or seek new E&R capabilities. Washington should still reach out to NSG partners that would be willing to sign on to a commitment.

The United States should also limit any proliferation potential of new technology it sells abroad to avoid undercutting an anti-E&R norm's future viability. For example, Washington, in cooperation with Urenco USA, is reviving its own uranium enrichment program to supply a near-20 percent enriched uranium fuel called high-assay low-enriched uranium, or "HALEU," for advanced reactors it plans to sell to foreign customers.[127] The majority of existing nuclear reactors worldwide run on low-enriched uranium and there is currently a "significant surplus" of enriched uranium worldwide, but only Russia currently supplies HALEU.[128] The development of an anti-E&R norm could be negatively impacted if Washington scales up commercial enrichment and shares near-highly enriched uranium (HEU) globally.[129] It must ensure, at a

---

127. The United States is planning several uranium enrichment sites for HALEU production. URENCO has operated its site in New Mexico since 2010. See: James Conca, "Supplying New Types of Nuclear Fuel with HALEU is Not a Problem," *Forbes,* September 8, 2021. (https://www.forbes.com/sites/jamesconca/2021/09/08/supplying-new-types-of-nuclear-fuel-with-haleu-is-not-a-problem/?sh=7106f19323be); Nuclear Innovation Alliance, "Advanced Nuclear Reactor Technology: A Primer," September 2021. (https://nuclearinnovationalliance.org/sites/default/files/2021-09/NIA_AdvancedNuclearReactorTech-Primer-Report.pdf)
128. World Nuclear Association, "Uranium Enrichment," Updated September 2020. (https://world-nuclear.org/information-library/nuclear-fuel-cycle/conversion-enrichment-and-fabrication/uranium-enrichment.aspx); John Funk, "DOE Targets End to U.S. Reliance on Russian Nuclear Fuel, Revived Domestic Capability," *Utility Dive*, July 20, 2020. (https://www.utilitydive.com/news/doe-targets-end-to-us-reliance-on-russian-nuclear-fuel-revived-domestic-ca/581701/)
129. See, for example: Victor Gilinsky and Henry Sokolski, "Dangerous Decisions about Advanced Reactors Could Lead to New Threats," *The National*

minimum, that reactor customers apply full-scope IAEA safeguards. U.S. administrations and Congress should meticulously evaluate the proliferation potential of such transfers.

The United States and its partners should continue to pursue Additional Protocol ratification and implementation by holdout countries. Should new states develop new E&R capabilities despite U.S. efforts, the IAEA must be better positioned to ensure the peacefulness of nuclear activities.

The NWS can help perpetuate an anti-E&R norm by carrying out further good-faith reductions of their nuclear arsenals. Some NNWS states, particularly members of the 120-member Non-Aligned Movement (NAM), often resist stronger nonproliferation commitments without corresponding progress on disarmament. They will likely refuse to sign the Additional Protocol or abide by an emerging E&R norm unless the NWS reduce numbers further. At first, they will also most certainly decry an E&R supply moratorium. Should security relations improve with Russia and China, the United States and Europe could, in the future, help matters by pursuing verified and reciprocal reductions with Moscow and Beijing which would support a norm-building initiative.

Conclusion

International norm building against E&R proliferation must entail consistent policy actions by a collective of states to solidify the broader global interpretation that spreading such capabilities runs counter to the NPT and threatens its nonproliferation aims. Measures taken since the NPT's signing support that most states

---

Interest, May 2, 2021. (https://nationalinterest.org/feature/dangerous-decisions-about-advanced-nuclear-reactors-could-lead-new-threats-183934)

desire greater E&R restrictions. This ongoing struggle will play out for years as Washington's adversaries, and even its partners, seek advanced fuel cycle technologies and restraint-favoring nations attempt to stop them. The United States can spearhead a norm against further E&R supply and development by announcing a new supply moratorium and building a coalition of likeminded states. Such actions support the widely shared objective of no further proliferation of nuclear weapons.

States must look to the overall intent of the NPT to prevent proliferation – even if it trumps what they view as an inalienable right to sensitive technology acquisition. The triumph of this interpretation is not worth the outcome. All countries must do their part to control the spread of nuclear weapons technology and ensure that additional states are not positioned a short dash to the bomb.

Reprocessing Constraints under the U.S.-China 123 Agreement

Eldon V.C. Greenberg

For a number of years, the Government of China has been interested in reprocessing spent nuclear fuel. China has a pilot reprocessing plant with a capacity of roughly 50 tons/year, and it is constructing a demonstration commercial reprocessing plant with a capacity of 200 tons/year. Further, since 2007, the French company, AREVA, has been in serious discussions with the Chinese to build a commercial-scale facility in Gansu that could process about 800-1000 tons/year, producing eight to ten tons of separated plutonium annually. According to the Nuclear Proliferation Assessment Statement for the U.S. agreement for nuclear cooperation with the Chinese government, developing a closed fuel cycle is a "near-term goal for China," and "China plans to reprocess most of its spent fuel domestically." *See* H. Doc. No. 114-28, 114th Cong. 1st Sess. 55, 56 (April 22, 2015). In the circumstances, how should the United States react to news that China was in fact proceeding to reprocess spent fuel from one of four, U.S.-supplied AP 1000 reactors? The answer turns particularly on the terms of the U.S. agreement for nuclear cooperation with the Chinese government. While that agreement does in general provide "advance consent" from the United States to Chinese reprocessing, it nonetheless gives the United States some modicum of control over actual Chinese reprocessing activities.

The Agreement Between the Government of the United States of America and the People's Republic of China Concerning Peaceful Uses of Nuclear Energy (the "China 123 Agreement") governs civil nuclear commerce between the United States and China. *See* House Doc. No. 114-28, 114th Cong., 1st Sess. (April 22, 2015). The China 123 Agreement was signed on April 13, 2015, and it was submitted to Congress on April 22, 2015. After lying before Congress for the requisite statutory review period under Section 123 of the Atomic Energy Act, 42 U.S.C. § 2153, the China 123 Agreement entered into force on October 29, 2015. It has a 30-year term and thus, unless terminated, will remain in force until 2045. *See* China 123 Agreement, Article 14, para. 2. The China 123 Agreement supersedes in all respects the prior, 1985 nuclear cooperation agreement with China, which was terminated on date the new agreement entered into force. It provides, "All the provisions of this Agreement shall apply to material, equipment, components, information and technology which were subject to the 1985 Agreement immediately prior to its termination." China 123 Agreement, Article 14, para. 4.

The China 123 Agreement contains a number of provisions relevant to Chinese reprocessing of "[n]uclear material transferred pursuant to . . . [the] Agreement and nuclear material used in or produced through the use of material or equipment so transferred." *See* China 123 Agreement, Article 6. In Article 6, para. 1, the China 123 Agreement establishes the general rule that there will be no reprocessing of covered material "unless the parties agree." In other words, the China 123 Agreement provides an overarching consent requirement for Chinese reprocessing of covered material. Thereafter, in Article 6, para. 2, the United States exercises the consent right provided for in Article 6, para. 1, for the term of the agreement. However, the advance consent given is subject to several important limitations also spelled out in the same paragraph. First, before the consent comes into effect, the parties "must agree on arrangements and procedures under which such reprocessing . . . shall take place." Second, reprocessing can only take place in facilities to which International Atomic Energy Agency ("IAEA") safeguards are applied. Third, if such facilities do not exist, and if the parties agree in writing, reprocessing may take place at facilities "eligible for IAEA safeguards" in accordance with China's 1988 agreement with the IAEA on the application of safeguards (IAEA INFCIRC/369). Fourth, relevant "arrangements and procedures" must include physical protection standards, storage standards and environmental standards consistent with other provisions of the China 123 Agreement. Fifth, separated nuclear material "may only be utilized at facilities to which the appropriate competent authorities of the Parties agree in writing."

Beyond Article 6, para. 2, certain other provisions of the China 123 Agreement are relevant to assessing the scope of U.S. control or influence over Chinese reprocessing. Thus, Article 7 specifies that "[a]dequate physical protection" must be maintained, consistent with and at least equivalent to the IAEA recommendations in IAEA INFCIRC/225 Rev. 5. Article 7, para. 4, further requires that the parties keep each other informed "in the event of unauthorized use or handling of nuclear material" covered by the agreement. In turn, Article 8 prohibits any explosive or military application of covered material. At the same time, Article 9, para. 1, requires that covered material in China "shall be subject" to the China-IAEA safeguards agreement of September 20, 1988, as well as the March 28, 2002, Additional Protocol to such agreement. If that agreement "is not being implemented," then Article 9, para. 3, requires the parties to establish "a mutually acceptable alternative."

Additionally, the China 123 Agreement contains a dispute resolution article (Article 12) which could come into play if the parties were in disagreement about application of any of the provisions discussed above. Article 12 calls for prompt consultations if concerns are raised about non-compliance. Importantly, it affirms that, should disputes arise, the parties "shall have the right to temporarily suspend or to cease further cooperation under the Agreement."

Finally, outside the text of the China 123 Agreement, there is an "Agreed Minute" which is declared to be "an integral part of the Agreement." It provides that, in the case of non-implementation of IAEA safeguards, the parties shall have the right to (1) review "in a timely fashion" equipment design; (2) to require "maintenance and production of records and of relevant reports;" and (3) to designate personnel to have "access to places and data necessary to account for" nuclear material, with an ability to inspect facilities, "install essential devices" and "make such independent measurements as may be deemed necessary to account for . . . nuclear material." The Agreed Minute, it should be noted, also applies to tritium production (which is subject to the peaceful use restrictions of Article 8), requiring annual reporting on its disposition to ensure that such disposition is consistent with Article 8.

Chinese reprocessing might, of course, have been subject to stricter oversight and control by the United States. In 2016, Senators Markey (D-MA) and Rubio (R-FL) introduced legislation which would have sharply enhanced the U.S. position if IAEA safeguards were not being applied. *See* S. 3010, 114th Cong., 2d Sess. (May 26, 2016). Section 9 of this bill would have prohibited the President from agreeing to any reprocessing at a facility to which IAEA safeguards were not being applied without certifying that a number of stringent conditions were being met, including, but not limited to, satisfaction of a "timely detection" standard for any diversion for military purposes, and meeting detailed reporting requirements. The Markey-Rubio bill, however, was not adopted by the Congress.

Even without the Markey-Rubio "fix", it is apparent that the United States is not without significant leverage under the China 123 Agreement. As noted above, despite the United States giving its advance consent to reprocessing, a series of agreements from the United States is required before China can proceed to reprocess covered material; physical protection must be provided; storage must be satisfactory; environmental standards must be met; military and explosive use must not occur; and safeguards must be applied. Article 9, para. 3, even contemplates that, if IAEA safeguards are not being implemented, a "mutually acceptable alternative" must be established, and the Agreed Minute gives the United States substantial power to ensure that the alternative is adequate. If there is a dispute, while consultation is required, suspension of cooperation is an option.

This is not to say that there are not ambiguities. For example, while application of China's safeguards agreement with the IAEA is central to the China 123 Agreement, China has the right under such agreement to remove facilities from safeguards and to veto inspectors. If it delisted its reprocessing facility, and decided to use the separated material for military purposes, at least as far as the IAEA is concerned China would still technically be in compliance with its obligations to the international agency. It might argue in this circumstance that its agreement with the IAEA is still being "implemented" and alternative safeguards arrangements, as contemplated by Article 9, para. 3, and the Agreed Minute, are not required. In response to efforts to establish such alternatives, it could also resist U.S. entreaties by claiming that such alternatives would "hamper, delay or [cause] undue interference" with China's nuclear program, in contravention of Article 9, para. 4, of the China 123 Agreement. While the arguments in support of this stance are a significant stretch, whether the China 123 Agreement would be sufficient to protect the United States' non-proliferation interests and provide the necessary latitude to challenge China's reprocessing activities remains to be seen.

The NPT's In Peril: How a New Article VI Agenda Might Save It

Thomas Grant

I. Introduction: A Tale of Two Treaties[130]

The Treaty on the Non-Proliferation of Nuclear Weapons[131]—the NPT—has passed its half-century mark as the cornerstone[132] of international efforts to curb the proliferation of nuclear weapons and reduce the risk of destabilizing arms races. With the conclusion of the tenth five-yearly NPT Review Conference in August 2022,[133] however, the NPT finds itself at risk. From two

---

130. With credit for the paraphrase "tale of two treaties" to Charles Dickens and the late James Crawford, Australian Judge of the International Court of Justice: James Crawford, "International Law and the Problem of Change: A Tale of Two Conventions," 49 *Victoria University of Wellington Law Review* 459 (2017).
131. Treaty on the Non-Proliferation of Nuclear Weapons (NPT), opened for signature July 1, 1968; entered into force March 5, 1970, 21 U.S.T. 483, 729 U.N.T.S. 161.
132. Reference to the NPT as the "cornerstone" of nuclear arms control is commonplace. See, e.g., Working paper submitted by New Zealand on behalf of the New Agenda Coalition (Brazil, Egypt, Ireland, Mexico, New Zealand, and South Africa), 15 March 2018, Preparatory Committee (Prep. Com.) for 2020 Review Conference (Rev. Con.): NPT/CONF/2020/PC.11/WP.13, para. 4.
133. In accordance with NPT Article VIII, paragraph 3, the NPT Parties held a conference five years after the NPT's entry into force "to review the operation of [the treaty] with a view to assuring that the purposes of the Preamble and the provisions of the Treaty are being realized." At five-yearly intervals thereafter, further Review Conferences were convened on majority vote of the Parties. In accordance with NPT Article X, paragraph 2, the NPT Parties convened and decided (in 1995) that the Treaty shall continue in force indefinitely: 1995 Rev. Con. Decision 3, NPT/CONF.1995/32 (Part I) p. 12 (May 12, 1995), https://documents-dds-ny.un.org/doc/UNDOC/GEN/N95/178/16/pdf/N9517816.pdf?OpenElement#page=17. The NPT Parties also convene a Preparatory Committee ("Prep. Com.") in each of the three years prior to a Rev. Con. See GA res. 47/52 A, December 9,

sides in the United States, stakeholders whose motivations would seem mutually exclusive converge in discontent over the NPT. Defense policy thinkers, especially conservatives, view the NPT's Article VI obligation to pursue negotiations at best with misgivings. For them, Article VI looks too much like the thin end of a wedge aimed to pry away nuclear deterrence from the national security strategy of the United States and the free world. Meanwhile, disarmament activists now view the NPT's negotiations requirement as a dead end. As they see it, the nuclear-weapon States[134] have too long neglected NPT Article VI and its more ambitious disarmament goal.

So much have these discontents with the NPT grown, albeit from sharply different premises, that we can fairly speak today of the NPT, and in particular its Article VI negotiation requirement, as a policy orphan.

At the international level, despite the United States's best efforts, negotiations have come to a standstill. The failure of negotiations to move forward places stress on the treaty. The NPT obliges the vast majority of its parties—the non-nuclear-weapon States—never to pursue or to have nuclear weapons; and the others—the five nuclear-weapon States—never to supply a nuclear weapon or the means to get one to a non-nuclear-weapon State. Though these provisions—under Article II and Article I of the NPT, respectively—are unconditional, many non-nuclear-weapon states

---

1992, https://documents-dds-ny.un.org/doc/UNDOC/GEN/N92/843/65/IMG/N9284365.pdf?OpenElement

134. "Nuclear-weapon State" is the expression that the NPT uses to refer to each State that "has manufactured or exploded a nuclear weapon or other nuclear explosive device prior to 1 January, 1967": NPT Article IV, paragraph 3. The nuclear-weapon States for purposes of the NPT therefore are China, France, Russia, the United Kingdom, and the United States. Other States, for purposes of the NPT, are referred to as "non-nuclear weapon States." The present paper uses the term "State" as it is used in the NPT to refer to an independent sovereign country recognized by other States to have capacity to hold rights and obligations under international law.

read them as part of a *quid pro quo*. On that reading, the nuclear-weapon states must implement Article VI in full, or else the constraints on the non-nuclear-weapon states fall away. The *quid pro quo* reading of the NPT is open to serious doubt as treaty interpretation, but it is widespread, even prevalent, among the States parties as a political position. According to that position, laxity toward disarmament justifies laxity toward nonproliferation. The NPT thus would seem no longer to promote the reduction of nuclear arms among the States that have them, nor to serve as a real check on others who might seek to get them. Little surprise, then, that across a wide spectrum activists and policy thinkers alike have begun to question whether the NPT is fit to serve the purpose its drafters intended.

However, while certainly not the best of treaties, the NPT is far from the worst. A new treaty, the Treaty on the Prohibition of Nuclear Weapons—the TPNW—pursues a very different approach toward disarmament from the NPT—and one that should lead security-minded policy thinkers to pause before we decide to retire the NPT. The TPNW, concluded in July 2017 and entered into force in January 2021,[135] in contrast to the NPT requires every state party that had a nuclear weapon after the TPNW's date of conclusion to carry out the "irreversible elimination" of nuclear weapons and nuclear-weapon programs.[136] No nuclear-weapon state has signed the TPNW or participated in its negotiation, but the TPNW already is attracting support in many countries that rely on the U.S. nuclear deterrent—including in countries on which the *United States* relies as trade partners and even as allies.

The difficulty with the TPNW is not that it reflects an aspiration to achieve a nuclear-weapon-free world. The difficulty with the

---

135. Treaty on the Prohibition of Nuclear Weapons (TPNW), concluded July 7, 2017; entered into force January 22, 2021: U.N.T.S. reg. no. 56487.
136. TPNW Article 4(1).

TPNW is that it confuses the aspiration with reality—and, worse, its unrealistic premise that the world is ready to disarm finds a ready audience at a time when the dwindling engagement of policy makers with NPT Article VI fosters the impression that the incrementalism of good faith negotiations has failed. The story of the arms race and disarmament, once told only in terms of the NPT, is now a tale of *two* treaties—the NPT and the TPNW. Even if we dispense with the superlative degree of comparison (the labeling of "bests" and "worsts"), the virtues and the faults of the former are to be appreciated in the sobering light which the coming to life of the latter now affords. In short, the less promising the negotiations provision of the NPT, the more attractive, at least to key constituencies in many countries, the fast-track disarmament premise of the TPNW.

Policy makers who care about strategic stability and security need to consider how the new treaty affects the old. China's breakneck buildup of nuclear arms makes clear that arms racing remains with us, and, thus, the arms race limb of NPT Article VI remains relevant today.[137] China's buildup also makes clear that rapid disarmament is no realistic option (if ever it was): unless and until the threat to stability and security wanes that China's buildup presents, we must navigate that geopolitical reality. But, if the TPNW, with its non-negotiable directive to disarm, continues to gain traction, then navigating the reality created by an unprovoked nuclear arms buildup will be at least as frustrating as it was at any time during the Cold War. If security-minded policy makers find

---

137. As to the arms race limb and its continued relevance, see Thomas D. Grant, *China's Nuclear Build-Up and Article VI NPT: Legal Text and Strategic Challenge* (Fairfax, Virginia: National Institute for Public Policy: Occasional Paper, Vol. 1, No. 11, November 2021) p. 40, https://nipp.org/wp-content/uploads/2021/11/Grant-OP-for-web.pdf#page=49; Thomas D. Grant, "Article VI of the Treaty on the Non-Proliferation of Nuclear Weapons (NPT): The Pursuit of Negotiations and China's Pursuit of Arms," 31 *Journal of Transnational Law & Policy* 1, 28-44 (2021-2022).

the negotiations clause of NPT Article VI an irritant and obstacle, then we will find the TPNW's absolute ban that much worse. The time has come to reinvigorate the NPT, in particular by engaging afresh with its Article VI negotiations clause. The stakes—in a world of burgeoning support for a prohibitionist alternative—are too high to remain ignored.

Following this introduction (Part I), the present paper addresses the emergence of the TPNW as a challenge to the NPT (Part II). It then outlines ten topics that the United States might consider advancing to refresh the negotiating agenda under Article VI of the NPT (Part III). A concluding section (Part IV) summarizes the main points and offers general reflections on how the NPT can supply the starting point for a pragmatic security diplomacy in service to our shared interest in stability and the reduction of nuclear risk.

II. The TPNW and the Absolutists' Challenge

The NPT for nearly fifty years was the only international legal instrument that embodied the aspiration to rid the world of nuclear weapons. The states that drafted the NPT recognized that the challenge is not to reach consensus about a goal; they already shared the goal of nuclear disarmament. The challenge was, and remains, instead, how to achieve that goal in a world where strategic realities constrain arms control. They drafted the NPT with the realities in mind.[138] Today, however, as Part I above suggested, we live in a world of *two* treaties concerned with

---

138. The NPT requires an approach that is both realistic and incremental, this being reflected in the preamble where the parties "[d]eclar[e] their intention to achieve at the earliest *possible* date the cessation of the nuclear arms race and to undertake effective measures in the *direction* of nuclear disarmament…" (emphasis added). As for the phrase "at an early date" in Article VI, this does not denote a timeline. See, e.g., Crawford, 49 *Victoria University of Wellington L. Rev.* at 459 (2017).

nuclear disarmament, not just one. The second treaty—the TPNW—now gathers adherents. More governments are signing the TPNW as time goes on. Of equal consequence or more, the disarmament community, which once placed its faith in the NPT, has shifted its focus to the TPNW. If we lived in an "echo chamber" in which realists, having little enthusiasm for the NPT, remained silent, and in which disarmament activists led the NPT discourse, then now we are approaching the point where even the echo is starting to fade. The activists now promote the TPNW and have less and less to say about the "cornerstone" that preceded it.

The subsections that follow consider (A) how disarmament absolutists met the TPNW with a groundswell of applause; (B) how the realists' resiling from the NPT joins with the disarmament community's disappointment to create a political vacuum that the TPNW is available to fill; and (C) how the TNPW, if it gains ascendance over the nuclear weapons discourse, will interfere with the security strategy of the United States and our democratic allies much more than our geopolitical competitors, China and Russia.

A. NPT and TPNW Hand in Hand—or the Absolutists' Triumph?

The disarmament community is at pains to reassure the countries whose accession to the TPNW they seek that the TPNW somehow complements the NPT and, in any event, in no way replaces it.[139]

139. According to UnfoldZero, the TPNW "supports the NPT and serves as part of NPT implementation by non-nuclear States":
https://www.unfoldzero.org/treaty-on-the-prohibition-of-nuclear-weapons/
SIPRI asserts that "the core obligations of the two treaties… seem to be perfectly consistent with one another" but also acknowledges that there is a "perceived incompatibility between the two," having "mainly… to do with the *indirect negative consequences* that the TPNW could potentially have on the NPT's non-proliferation objectives." Dr Tytti Erästö, The NPT and the TPNW:

There are states that say so too. For example, members of the New Agenda Coalition said in connection with the 2020 NPT Review Conference[140] that the TPNW "complements and strengthens" the NPT "as the cornerstone of the nuclear disarmament and non-proliferation regime."[141] So the proposition that the TPNW does not displace the NPT, but functions *with* it hand in hand, is gaining currency at governmental level.

Yet, on inspection, to propose that the NPT and TPNW are destined to co-exist is dubious. In the years immediately after the ICJ's Advisory Opinion on *Nuclear Weapons,* the disarmament community took encouragement from the ICJ's interpretation of Article VI of the NPT; disarmament activists were hopeful that the Court's assertion that Article VI obliges disarmament as a *result,* as distinct from obliging that States *endeavor* toward that result, had opened the door to rapid disarmament. The activists' hopes

---

Compatible or conflicting nuclear weapons treaties?" SIPRI WritePeace blog (March 6, 2019), https://www.sipri.org/commentary/blog/2019/npt-and-tpnw-compatible-or-conflicting-nuclear-weapons-treaties See also Müller & Wunderlich, 149(2) *Daedalus* at 172: "Reasonable policies can create a *modus vivendi* [between the NPT and TPNW]."

140. In accordance with NPT Article VIII, paragraph 3, the NPT Parties held a conference five years after the NPT's entry into force "to review the operation of [the treaty] with a view to assuring that the purposes of the Preamble and the provisions of the Treaty are being realized." At five-yearly intervals thereafter, further Review Conferences were convened on majority vote of the Parties. In accordance with NPT Article X, paragraph 2, the NPT Parties convened and decided (in 1995) that the Treaty shall continue in force indefinitely: 1995 Rev. Con. Decision 3, NPT/CONF.1995/32 (Part I) p. 12 (May 12, 1995), https://documents-dds-ny.un.org/doc/UNDOC/GEN/N95/178/16/pdf/N9517816.pdf?OpenElement#page=17. The NPT Parties also convene a Preparatory Committee ("Prep. Com.") in each of the three years prior to a Rev. Con. See GA res. 47/52 A, December 9, 1992, https://documents-dds-ny.un.org/doc/UNDOC/GEN/N92/843/65/IMG/N9284365.pdf?OpenElement

141. See, e.g., Working paper submitted by New Zealand on behalf of the New Agenda Coalition (Brazil, Egypt, Ireland, Mexico, New Zealand, and South Africa), 15 March 2018, Prep. Com. for 2020 Rev. Con.: NPT/CONF/2020/PC.11/WP.13, para. 4.

dwindled over time.[142] The ICJ's refusal to address the merits of recent claims by the Marshall Islands against the nuclear-weapon States under Article VI[143] probably brings that phase of disarmament aspiration to a close.

It is against the backdrop of their disappointment in Article VI that the disarmament community celebrated the arrival of the TPNW. Among the encomia, writers described the TPNW as "a monumental achievement" and a "tipping point" bringing to fruition "widespread agreement" on disarmament;[144] a "turning point,"[145] a "major milestone in international law and nuclear security policy,"[146] and "one of the most important diplomatic developments of the modern era."[147] A note of triumphalism entered the disarmament discourse.

---

142. For a psychologizing assessment of the disarmament activists' faded hopes and a resultant search for "emotional satisfaction," see Müller & Wunderlich149(2) *Daedalus* at 177 (2020).
143. See in particular *Obligations concerning Negotiations relating to Cessation of the Nuclear Arms Race and to Nuclear Disarmament (Marshall Islands* v. *United Kingdom),* The Marshall Islands asked the ICJ to hold that the UK had "violated and continues to violate its international obligations under the NPT, more specifically under Article VI of the Treaty, by failing to pursue in good faith and bring to a conclusion negotiations leading to nuclear disarmament…" Judgment, October 5, 2016, I.C.J. Rep. 2016 p. 833, 839, para. 11: https://icj-cij.org/public/files/case-related/160/160-20161005-JUD-01-00-EN.pdf#page=10 The ICJ declined to reach the merits of the Marshall Islands' request, on the ground that the Marshall Islands had not identified a dispute with the United Kingdom subject to the ICJ's jurisdiction: I.C.J. Rep. 2016 at p. 856 para. 58.
144. Claire Slatter, "Securing the Pacific in a Globalised World: New and Emerging Developments in International Law," 27 *Canterbury Law Review* 5, 6-7 (2020).
145. Daryl G. Kimball, "A Turning Point in the Struggle Against the Bomb: The Nuclear Ban Treaty Ready to Go Into Effect," *Just Security* (Oct. 27, 2020), https://www.justsecurity.org/73050/a-turning-point-in-the-struggle-against-the-bomb-thenuclear-ban-treaty-ready-to-go-into-effect/
146. David A. Koplow & Edward M. Ifft, "Legal and political myths of the Treaty on the Prohibition of Nuclear Weapons, *Bulletin of the Atomic Scientists* (May 13, 2021).
147. Kennedy Graham, "The TPNW Conference of Parties: What Is to Be Discussed?" 3 *Journal for Peace and Nuclear Disarmament* 234 (2020).

B. Nonproliferation Abhors a Vacuum

Security-minded policy makers might ask, if the NPT and its associated negotiating agenda were to fade into disuse, then, so what? Surely, the important thing is that we maintain our deterrent. There are deficiencies, however, in a "so what" response.

For one, the deterrent we need is shaped by the security environment in which we operate, and nonproliferation shapes that environment. When nonproliferation *works,* it shapes the security environment for the better. The United States and our allies, while we must be ready for a quantitative nuclear arms race, benefit when we shape strategic competition in ways that favor our other strengths and capabilities. Disappearance of nonproliferation as an operative category of security policy does not equate to a security benefit in a world in which malign actors continue to seek nuclear weapons and those who already have them are emboldened to threaten to use them as an umbrella for conventional armed aggression.[148]

---

148. The first branch of this observation is straightforward: the decline of nonproliferation diplomacy hinders efforts to keep nuclear weapons out of the hands of malign actors (whether States or others). The second branch—observing that a nuclear-weapon State (Russia) is conducting conventional armed aggression against non-nuclear-weapon States and that this situation, too, recommends maintaining a viable nonproliferation diplomacy—merits elaboration: conventional armed aggression conducted by a nuclear-weapon State that uses the threat of nuclear escalation to dampen the defender's response (and that of the defender's supporters) is a potent incentive to proliferate, because it communicates to non-nuclear-weapon States that they are vulnerable to attack. In short, one reason to preserve our nonproliferation diplomacy has not gone away—malign would-be proliferators; and a new reason arises from Russia's aggression—non-nuclear weapon States justifiably seeking the means to assure their own self-defence.

See also the slightly different observation, by U.S. Secretary of Defense Austin at the Halifax International Security Forum, that Russia's aggression gives a proliferation incentive to would-be aggressor-States, because they "could well conclude that getting nuclear weapons would give them a hunting license of their own." U.S. Department of Defense, "Austin Says Ukraine Matters to all

Moreover, disappearance of a realistic nonproliferation agenda would not leave a hole in this policy space forever. If the NPT were to fade, then this would create a vacuum, and nonproliferation abhors a vacuum. We now know what would fill the vacuum: the TPNW is the matter waiting to rush in.

Uptake of the TPNW in the first two years after it opened for signature was rapid.[149] At least some of the states' parties to the TPNW appear to be taking seriously the obligation under Article 12 TPNW to "encourage States not party to this Treaty to sign, ratify, accept, approve or accede to the Treaty, with the goal of universal adherence of all States to the Treaty." [150] As of November 2022, the TPNW has 93 signatories, of which 68 are parties.[151] It is too early to say that the TPNW has overtaken the arms control and disarmament agenda; just as it is too early to say that the NPT has declined beyond recovery. However, the trends are in that direction.

The NPT and the TPNW overlap in a crucial way. Their overlap is between the NPT's negotiation provision and the TPNW's

---

Peoples and Nations" (November 19, 2022): https://www.defense.gov/News/News-Stories/Article/Article/3224096/austin-says-ukraine-matters-to-all-peoples-and-nations/ As to continuity on this point between the Trump Administration and the Biden Administration, with intellectual groundwork found earlier still, see Ford (Jan. 2023), op. cit, p. 9.

149. See *United Nations Disarmament Yearbook 2019: Part II,* p. 246. A glance at the chart in the *Yearbook* might suggest that the TPNW is poised to overtake the CWC, NPT, BWC, and CTBT in number of participants, but the chart might confuse an incautious reader: it shows the rate at which new participants joined the TPNW in the *first two years* after the TPNW opened for signature and compares that steep curve to the much shallower one depicting the rate at which new participants had joined the four older arms control treaties *decades after* they had opened for signature. The rhetorical effect is noteworthy nevertheless: it suggests that the TPNW is a treaty on the move, the others at a plateau.

150. See Nuclear Weapons Ban Monitor, "The obligation to promote universal adherence to the Treaty," https://banmonitor.org/positive-obligations-1/the-obligation-to-promote-universality-of-the-treaty

151. United Nations Office for Disarmament Affairs, Status of the TPNW: https://treaties.unoda.org/t/tpnw

prohibitionary obligation. The NPT pledges its parties to negotiate toward arms control and disarmament; the TPNW obliges them to disarm. This is the difference between a treaty that nuclear-weapon States painstakingly crafted to address the world in which they existed—a world the lineaments of which are very much visible still today—and an aspirational exercise calling for radical change. The problem with the aspirational exercise is that it pledges countries to implement a disarmament end-state without regard to the risks that that end-state would entail if only *some* countries reached it; and without regard to the effort required if *all* countries with nuclear weapons are to agree to effective disarmament measures.

The 2022 NPR reiterates that the TPNW fails to serve the interests of international security. According to the 2022 NPR,

> While the United States actively pursues the goal of a world without nuclear weapons, it does not consider the Treaty on the Prohibition of Nuclear Weapons (TPNW) to be an effective means to reach that goal. The United States does not share the underlying assumption of the TPNW that the elimination of nuclear weapons can be achieved irrespective of the prevailing international security environment. Nor do we consider the TPNW to be an effective tool to resolve the underlying security conflicts that lead states to retain or seek nuclear weapons.[152]

This is a welcome statement. However, it might have gone further.

As expressed in the 2022 NPR, the United States' concerns about the TPNW appear to be chiefly *negative*: the TPNW does not do what we need a nonproliferation policy to do. The deficiency in expressing our concerns in this limited and negative way is that TPNW's failings are not limited to what it does not do. The TPNW

152. 2022 NPR at p. 19.

fails much more seriously in what it *does* do: it gives the false impression that a declaration in legal form on its own can expedite the attainment of security through disarmament. In giving that false impression, the TPNW motivates civil society in a range of countries to pursue shifts in nuclear policy that not only will fail to attain security but also run the risk of undermining strategic stability.

Before turning to how we might re-invigorate NPT Article VI, and, by so doing, prevent a damaging drift of allies and partners toward the TPNW, we need to remind ourselves of the full measure of the challenge: the TPNW affects different countries in different ways. A few words are in order about the asymmetric way in which the TPNW affects democratic countries.

C. The TPNW's Asymmetric Effects

From early in the study of nuclear deterrence, scholars have concerned themselves with the asymmetric impact of different possible arms control initiatives. Classic treatments of arms control asymmetry were concerned, chiefly, with weapons—how reductions in particular categories of weapons might have asymmetric impact as between the Cold War "power blocs."[153] However, a given arms control initiative also may have asymmetric *political or rhetorical* impact. An authoritarian, one-party state that allows little room for public discourse on arms control and, in any event, is little influenced by public preferences in regard to arms control, feels the effects of an arms control initiative very differently from a democratic state with an open society.

---

153. See Schelling & Halperin (1961) at p. 54.

William Van Cleave, who advised the Reagan Administration on nuclear arms control and, among other roles, served on the U.S. Delegation to the Strategic Arms Limitation Talks, observed nearly forty years ago that

> in the democratic states of the West there is *always* arms control, even without negotiated agreements. Arms are controlled and limited by the West's traditional values, by its political and budgeting process, and by the influence of the media and of public opinion.[154]

We disregard this axiom—that, in our societies, there "is *always* arms control"—at our peril. Its salience today has grown as the disarmament discourse has shifted from Article VI of the NPT to the absolutist prohibition of the TPNW.

It is all too easy to conclude that the TPNW, because its goals are unrealistic, will not affect us. The seemingly settled assumption that we *will* continue a strategically-minded policy of nuclear deterrence, because disarmament is simply unworkable in present circumstances, lures too many in the policy community into complacence.

Also luring us to stop worrying and ignore the TPNW is that all the nuclear-weapon states reject the TPNW. According to the P5 Joint Statement on the TPNW (October 24, 2018), "The TPNW will not be binding on our countries, and we do not accept any claim that it contributes to the development of customary international law; nor does it set any new standards or norms."[155] China, in addition to

---

154. William R. Van Cleave, "The Arms Control Record: Successes and Failures" chapter in Richard F. Staar, ed., *Arms Control: Myth Versus Reality* (Stanford University Press, 1984) p. 3 (emphasis original).
155. P5 Joint Statement on the TPNW (October 24, 2018), available at https://www.gov.uk/government/news/p5-joint-statement-on-the-treaty-on-the-nonproliferation-of-nuclear-weapons. As to the non-effect of the TPNW on customary international law, see Thomas D. Grant, "The ILC's Draft

joining the P5 Joint Statement rejecting the TPNW, separately "emphasized that the [NPT] is the cornerstone of the international nuclear non-proliferation system and an important component of the international security architecture."[156] According to China, "China respects the role of the Non-Proliferation Treaty as the cornerstone of the international nuclear non-proliferation regime."[157] But civil society in China (and in Russia) has little or no influence over nuclear weapons strategy. The effect of the TPNW is asymmetric, even as the five nuclear-weapon states seem to have a shared interest in tempering the absolutism behind it. A degree of caution is in order, therefore, when we consider the stated opposition of Russia and China to the TPNW. Even if they are sincere, they are not affected the same way we are by the suasive influence of disarmament discourse.

We also need to consider how disarmament discourse affects like-minded countries, including our closest allies. If the TPNW comes to occupy the political space that heretofore belonged to the NPT, then serious challenges will arise for our security diplomacy. Public affirmations by allies that it is still the NPT that matters, though no doubt sincere, should no more comfort us than affirmations to the same effect by China or Russia. For example, countries comprising the Vienna Group of Ten stated in the Preparatory Committee leading to the NPT 2020 Review Conference that it is still the NPT that matters. They reiterated their "full commitment" to the NPT, which they affirmed is "*the* cornerstone of the international nuclear disarmament and non-

---

Conclusions on Customary International Law and the Treaty on the Prohibition of Nuclear Weapons (TPNW): Lawmaking and its Limits in a Nuclear Age," 13(2) *George Mason International Law Journal* 1-53 (2022).

156. NPT/CONF.2020/PC.III/8 (April 29, 2019), Implementation of the Treaty on the Non-Proliferation of Nuclear Weapons of the People's Republic of China, Report submitted by China, para. 31.

157. *Id.* at para. 79.

proliferation regime" (emphasis added).[158] But, while the governments of China and Russia have justified confidence in their ability to resist public pressure to move toward the TPNW's prohibition (there will be little if any public pressure on them), and so their affirmations of the NPT are easy to make and easy to *continue* to make, the governments of U.S. allies and like-minded countries must follow the will of the electorates who choose them. If a political vacuum arises around arms control in democratic countries and the TPNW fills it, then the security policy of democratic countries will change accordingly.

A shift in the free world toward the TPNW is no mere conjecture. Australia, under a coalition government comprising the Liberal Party and National Party from August 2018 through May 2022, voted repeatedly against the TPNW in the UN General Assembly.[159] The coalition government, after the federal election of May 21, 2022, was replaced by a Labor Party government. Australia, on October 28, 2022, shifted position: instead of casting a negative vote against the TPNW in the General Assembly, it abstained.[160] Disarmament groups noted the shift with approval.[161]

---

158. NPT/CONF.2020/PC.III/WP.2 (March 15, 2017), Addressing "Vienna issues", Working paper submitted by Australia, Austria, Canada, Denmark, Finland, Hungary, Ireland, the Netherlands, New Zealand, Norway and Sweden (the Vienna Group of Ten), para. 1.
159. See, e.g., GA res. 75/40, Dec. 7, 2020, and for voting record see A/75/PV.37 (Dec. 7, 2020) p. 15: https://documents-dds-ny.un.org/doc/UNDOC/GEN/N20/343/94/PDF/N2034394.pdf?OpenElement#page=15
160. For the October 28, 2022 GA resolution in draft, see https://reachingcriticalwill.org/images/documents/Disarmament-fora/1com/1com22/resolutions/L17.pdf
161. See, e.g., "States reaffirm support for Nuclear Weapons Ban Treaty at UN First Committee," *ICAN Updates* (October 29, 2022): https://www.icanw.org/states_reaffirm_support_for_nuclear_weapons_ban_treaty_at_un_first_committee

The TPNW also continues to receive support among states in the Non-Aligned Movement.[162]

If the TPNW gains further adherents, then it might well offset, even negate, diplomatic advantages that are emerging from other trends. For example, commentators have suggested that China's failure to condemn Russia for the latter's war of aggression against Ukraine has eroded willingness among governments in Eastern and Central Europe to transact with China.[163] While no doubt Russia's aggression will give pause to countries in that region before they adopt a prohibitionist stance against nuclear weapons—consider, for example, Finland's suggestion that it might one day host nuclear weapons[164]—the TPNW, if it continues to attract support in civil society, pushes the other way. No symmetric effect arises from the TPNW in the authoritarian systems that prevail in China and Russia.

III. What's Left to Negotiate?

This paper has suggested that we take a fresh look at NPT Article VI, with the aim of re-invigorating the negotiated approach to arms control that that long-standing treaty provision calls for. However,

---

162. See Final Document of the 18th Non-Aligned Movement Summit of Heads of State and Government, Baku (October 25-26, 2019), NAM 2019/CoB/Doc.1, p. 81 para. 245: the Non-Aligned Movement Heads of State and Government "*took note* of the adoption of the [TPNW]…" which, they "hoped" will "contribute to furthering the objective of the total elimination of nuclear weapons."

163. Hiroyuki Akita, "How China continues to lose friends in Central and Eastern Europe," *Nikkei Asia* (November 3, 2022): https://asia.nikkei.com/Politics/International-relations/How-China-continues-to-lose-friends-in-Central-and-Eastern-Europe

164. Leo Laikola & Niclas Rolander, "Nordic NATO Candidates Seek No Opt-Outs on Nuclear Weapons," *Bloomberg* (November 1, 2022): https://www.bloomberg.com/news/articles/2022-11-01/nordic-nato-candidates-seek-no-opt-outs-on-nuclear-weapons

if negotiating is to address the risk that the existence of nuclear arsenals entails, then we must identify meaningful issues to negotiate about.

In an Occasional Paper published in June 2022, the Nonproliferation Policy Education Center suggested a number of matters that the United States might invite China to address in negotiations. Among the possibilities suggested in that paper are restricting plutonium production, establishing nuclear hotlines, clarifying the terms of the Comprehensive Test Ban Treaty, and adopting new transparency measures, such as IAEA safeguarding of the most "military useful" but ostensibly "civilian" nuclear facilities.[165] A table top exercise that NPEC conducted in 2022 reached the conclusion that non-aligned countries would not be likely to support U.S. efforts to place pressure on China to enter NPT Article VI negotiations, if the United States did not propose any serious steps toward nuclear limitation.[166]

To advance a new negotiations agenda, the United States needs to take care in selecting the subject matter that we prioritize. Policy grounds exist, for example, to pursue expanded homeland missile defense. Such defense increases strategic stability and, so, furthers a central NPT aim.[167] Moreover, if history is any guide, vigorous pursuit of missile defense by no means precludes arms control: the United States pursued missile defense in the final years of the Cold War, and that period coincided with major breakthroughs in U.S.-Soviet arms control negotiations, including the Intermediate Range

---

165. Henry Sokolski & Andrea Beck (eds.), *Arresting Nuclear Adventurism: China, Article VI, and the NPT* (Nonproliferation Policy Education Center: Occasional Paper 2201, June 2022) 5-6.
166. NPEC Occasional Paper, p. 5.
167. See Matthew R. Costlow, *Vulnerability is No Virtue and Defense is No Vice: The Strategic Benefits of Expanded U.S. Homeland Missile Defense* (National Institute for Public Policy, Occasional Paper Vol. 2, No. 9, September 2022): https://twitter.com/Matt_Costlow/status/1588539838332284928/photo/1

Nuclear Forces Treaty (INF) of 1987.[168] The United States today should pursue negotiations on subject matter that holds promise for advancing security, not on subject matter that erodes it.

Preserving strategic space for efforts such as missile defense does not foreclose other topics for negotiation. NPT Article VI both accommodates and encourages fresh approaches, including the exploration of new negotiation subject matter.[169] In view of the wide ambit of Article VI, the United States should consider exploring a number of substantive issues, so far largely neglected, that could comprise a new agenda for negotiations.

The substantive issues may be grouped under the following rubrics, each of which this section will consider in turn: *(A)* transparency; *(B)* confidence-building; *(C)* hotlines and other working-level arrangements; *(D)* plutonium production and fuel-cycle accountability; *(E)* safeguards for fast reactors; *(F)* nuclear testing and definitions in the CTBT framework; *(G)* nuclear safety; *(H)* export controls; *(I)* institutionalizing the NPT; and *(J)* the conventional-nuclear deterrence matrix. Overlaps and connections exist among these. For example, confidence-building is naturally linked to hotlines. Fuel-cycle accountability overlaps with safeguards agreements. Transparency has relevance to most or all the others. These ten rubrics nevertheless furnish a convenient starting point for the future negotiating agenda that this paper argues we need, if we are to breathe new life into the NPT and retain a viable security diplomacy at a time when a destabilizing alternative exerts a widening influence.

---

168. Costlow (September 2022), p. 50.

169. The wide ambit for potential negotiations subject matter is reflected in the language and structure of the NPT. See 31 *Journal of Transnational Law & Policy* at 24-27; National Institute for Public Policy Occasional Paper, volume 1, number 11 (November 2021) 39.

(A) Transparency

Policy makers long have recognized the role of transparency in arms control.[170] The 2022 NPR refers to transparency in connection with adversaries' nuclear activities.[171] It notes in particular China's "lack of transparency" in regard to its "nuclear expansion."[172] However, while referring to efforts to "enhance transparency" among the P5, the 2022 NPR does not connect those efforts directly to a renewed NPT Article VI agenda. A number of nonaligned countries and others, at the 2020 Review Conference, did make that connection, expressing the view that transparency is an important component of future efforts under the NPT.[173] The United States would gain diplomatic capital if we were to pursue negotiations in regard to transparency in the Article VI frame. Moreover, actual progress toward transparency from China would lend stability to a relationship beset today by increasing uncertainty.

---

170. See, e.g., the 1998 United Kingdom Strategic Defence Review (SDR), para. 61, stating that "[t]ransparency about nuclear weapons holdings… plays a part in arms control", quoted in *R. (on the application of Marchiori)* v *Environment Agency* [2001] EWHC Admin 267 (March 29, 2001; Mr Justice Turner); the U.S. Senate Resolution of Ratification of the Start II Treaty (January 26, 1996), part c(1) on cooperative threat reductions, reprinted at 90 *American Journal of International Law* 272, 275 (1996) (Marian Nash (Leich), ed.).

171. 2022 NPR at 13, 16, 18

172. 2022 NPR at 17.

173. See, e.g., Working paper submitted by New Zealand on behalf of the New Agenda Coalition (Brazil, Egypt, Ireland, Mexico, New Zealand, and South Africa), March 15, 2018, Prep. Com. for 2020 Rev. Con.: NPT/CONF/2020/PC.11/WP.13, para. 9, referring to Action 5 of 22 actions enumerated in the 2010 Rev. Con. action plan: "further enhancing transparency and increasing mutual confidence."

(B) Confidence-building

As Henry Sokolski observed in 2015, the pursuit of confidence-building measures is typically associated with arms control officials who search for near-term activities to fill a diplomatic agenda.[174] In contrast as Sokolski observed, "hawkish supporters of nuclear weapons (as well as hard-headed security planners who might not be as enthusiastic about relying heavily on nuclear arms)" focus on longer-term and seemingly more concrete goals.[175] Confidence-building has a formalistic ring about it that excites little support outside the foreign policy bureaucracy.[176]

And, yet, present-day circumstances give rise to new risks that confidence-building well might help ameliorate. China's rapid nuclear buildup, as well as the increase in the number of countries with nuclear fuel capabilities (about which more will be said below), have imparted instability and unpredictability to relations among nuclear-weapon states (and among states that might pursue nuclear weapons). So too has the eruption of conventional armed conflict involving threats of nuclear force—i.e., Russia's aggression against Ukraine—and the possibility of other conventional armed conflict—e.g., a future invasion of Taiwan by China.[177] Russia's aggression has violated the fundamental

---

174. Henry Sokolski, *Underestimated: Our Not So Peaceful Nuclear Future* (Arlington: Nonproliferation Policy Education Center, 2015) p. 5 n. 8.
175. *Id.*
176. For an exception, see William M. Moon, "Beyond Arms Control: Cooperative Nuclear Weapons Reductions—a New Paradigm to Roll Back Nuclear Weapons and Increase Security and Stability," 3(1) *Journal for Peace and Nuclear Disarmament* 92-114 (2020), in which the author proposes confidence-building measures as part of a possible future renewal of cooperative efforts with Russia. Moon, now a consultant to the Stimson Center, served, *inter alia,* as EUCOM Theater Support Team Lead for the Defense Threat Reduction Agency (DTRA).
177. China's aggressive acts against India suggest that China harbors territorial aims in that direction as well. See Manoj Joshi, *Understanding the India China Border. The Enduring Threat of War in the High Himalayas* (Hurst, 2022).

international law rule of the post-1945 era: the prohibition of use or threat of force for purposes of territorial aggrandizement.[178] It is to be hoped that others do not take the invasion of Ukraine as a cue to pursue their own irredentist projects, or as a warning that only states with nuclear weapons are safe; it remains too early to tell whether Russia's war-making is to be a precedent or an anomaly.

Present geopolitical unrest could be at least as dangerous as the Cold War stand-off between the superpowers. For much, if not all, of the Cold War, the United States and the Soviet Union each understood with relative confidence what capabilities the other possessed and what limits the other would respect. The superpowers developed informal understandings with one another, a practice that reflected and strengthened mutual restraint. (The practice drew remark from Schelling and Halperin at the time).[179] Through confidence-building measures, geopolitical competitors today could refurbish something akin to the informal understandings that lent stability to superpower relations in the past. The strategic environment today is one in which confidence-building measures, sensibly deployed, need not be mere time-fillers for a diplomat's diary but practical steps toward risk-reduction. Because rapid changes now are taking place in hard-to-predict directions, confidence-building is a tool that deserves a fresh look in the NPT negotiating frame.[180]

---

178. Regarding the stabilizing effect of the post-1945 territorial settlement, which long was a reality in geopolitics and in international law, see Thomas D. Grant, *Aggression Against Ukraine: Territory, Responsibility, and International Law* (Macmillan, 2015) pp. 101-167. The view that the prohibition against territorial change through threat or use of force is the principal post-1945 legal rule has gained currency since Russia escalated its aggression against Ukraine in February 2022. See, e.g., Ingrid (Wuerth) Brunk & Monica Hakimi, *Russia, Ukraine, and the Future World Order,* 116(4) AM. J. INT'L L. 687-697 (2022).
179. Thomas D. Schelling & Morton H. Halperin, *Strategy and Arms Control* (New York: The Twentieth Century Fund: 1961) 77-90.
180. The proposal that strategic instability recommends confidence-building relates to the analogous insight that a rush to nuclear arms *reductions* might well

(C) Hotlines

The United States and the Soviet Union established a hotline in 1963 which has continued to operate ever since.[181] The United Kingdom and France later in the 1960s established similar facilities with the Soviet Union.[182] India and Pakistan established hotlines with one another in the 1990s.[183] The NPEC diplomatic simulation conducted in autumn 2021 (and noted in this paper above) concluded that the United States should pursue, among other initiatives, improving nuclear crisis communications with

---

introduce risk if not conducted in a "transparent, coordinated fashion": Sokolski (2015) p. 5. Rapid change—even in a seemingly benign direction—upsets settled expectations and threat-perceptions.

181. Steven E. Miller, "Nuclear Hotlines: Origins, Evolution, Applications," 4 *Journal for Peace and Nuclear Disarmament* 176, 182 (2021). The United States and Soviet Union established the hotline in accordance with an MOU—the Memorandum of Understanding Regarding the Establishment of a Direct Communications Link (June 20, 1963): 472 U.N.T.S. 163, https://treaties.un.org/doc/Publication/UNTS/Volume%20472/volume-472-I-6839-English.pdf The hotline memorandum has been updated in subsequent agreements. See Agreement of September 30, 1971: 806 U.N.T.S. 402; Agreement of July 17, 1984: 2193 U.N.T.S. 51.

182. The United Kingdom and the Soviet Union concluded a hotline agreement in 1967. Agreement concerning the establishment of a direct communication link between the Residence of the Prime Minister of the United Kingdom in London and the Kremlin (August 25, 1967), 632 U.N.T.S. 49; Agreement on the Improvement of the Direct Communications Link between the Residence of the Prime Minister of the United Kingdom in London and the Kremlin (March 31, 1987), 1655 U.N.T.S. 393.

France and the USSR under a Joint Declaration of June 30, 1966 agreed to establish "a direct line of communication" in regard, *inter alia,* to nuclear weapons issues: noted at Jean Charpentier (ed.), "Pratique française du Droit International," 12 *Annuaire français de droit international* 919 (1966).

183. See Neil Joeck, "The Indo-Pakistani Nuclear Confrontation: Lessons from the Past, Contingencies for the Future," Nonproliferation Policy Education Center Working Paper Series (September 2008) p. 22: https://npolicy.org/wp-content/uploads/2008/09/The_Indo-Pakistani_Nuclear_Confrontation-1.pdf#page=22

China and should encourage China to do the same with its regional neighbors.[184]

A U.S.-China presidential hotline has been in place since 1998.[185] A Defense Telephone Link (DTL) was established in 2008 to connect the two countries' defense secretaries.[186] The U.S. Department of Defense at present calls for more U.S.-China hotlines,[187] and "several former senior U.S. military personnel" have been reported to say that the "inadequacy [of U.S. China crisis communications systems] constitutes a clear and present danger… that could fuel a dangerous U.S.-China military confrontation at a time of heightening bilateral tensions in the Taiwan Strait and the South China Sea."[188]

Hotlines are not a cure-all, and to use them effectively the countries that establish them must use them in good faith. Pakistan, for example, has complained of misuse of the hotlines that it

---

184. Henry Sokolski & Andrea Beck (eds.), *Arresting Nuclear Adventurism: China, Article VI, and the NPT* (Nonproliferation Policy Education Center: Occasional Paper 2201, June 2022) p. 13.
185. See Joint Press Conference, Secretary of State Madeleine K. Albright and Foreign Minister Tang Jiaxuan, April 29, 1998: https://1997-2001.state.gov/statements/1998/980429.html As to the long process that led to agreement to set up the U.S.-China hotline, see Miller (2021) at p. 183 n. 21. The process started in 1971 with the Nixon-Kissinger diplomatic opening with China. *Id.*
186. See Agreement on the Establishment of a Secure Defense Telephone Link between the Department of Defense, the United States of America and the Ministry of National Defense, the People's Republic of China (February 29, 2008): https://www.state.gov/wp-content/uploads/2019/02/08-229-China-Telecommunication-Link.EnglishOCR.pdf
187. Jack Detsch, "Biden Looks for Defense Hotline With China," *Foreign Policy* (May 10, 2021): https://foreignpolicy.com/2021/05/10/biden-china-xi-jinping-defense-hotline-pentagon/
188. Phelim Kine, " 'Spiral into crisis' The U.S.-China military hotline is dangerously broken," *Politico* (September 1, 2021): https://www.politico.com/news/2021/09/01/us-china-military-hotline-508140

established with India in the 1990s.[189] Nuclear weapons policy analysts have argued that misuse is an ever-present risk,[190] because gaps exist between hotline users' intentions. A recent RAND blog commentary cautions against overreliance on hotlines with China, because China intends to misuse them. According to that commentary, China intends hotlines, not to "deconflict and de-escalate... between forces during a crisis" but, instead, to allow China to "exploit[...] crises to its advantage and manipulat[e] risk calculations." If that view is right, then China will not use hotlines to resolve crises, but "to signal resolve, assign blame, and stall until Beijing stakes out a position of maximum pressure and leverage over the United States during negotiations."[191]

With each of the suggestions that the present paper makes, U.S. policy makers—and if it comes to a negotiation with China, U.S. negotiators—must exercise caution. If it does come to a negotiation on hotlines, then the United States should make clear that we intend any future hotline or updating of existing hotlines to further the larger objectives to which we long have been committed under NPT Article VI. Space nevertheless exists to consider hotlines among possible topics for a new negotiating agenda.

---

189. See General (ret.) Feroz Khan, "Prospects for Indian and Pakistani Arms Control and CBMs," Nonproliferation Policy Education Center (February 24, 2010) p. 14: https://npolicy.org/wp-content/uploads/2021/08/Prospects_of_Arms_Control_and_CBMS_Between_India_and_Pakistan-ESSAY.pdf#page=14 Cf. regarding underutilization of the India-Pakistan hotlines, Joeck, op cit, at p. 35.
190. See Miller (2021) at p. 181.
191. Lyle J. Morris & Kyle Marcrum, "Another 'Hotline' With China Isn't the Answer," TheRANDblog (July 27, 2022): https://www.rand.org/blog/2022/07/another-hotline-with-china-isnt-the-answer.html

(D) Plutonium Production and Fuel-cycle Accountability

China in 2017 discontinued its earlier practice of reporting to the IAEA on its plutonium holdings.[192] Coinciding with that retrograde step, China started work on nuclear fuel reprocessing facilities to extract plutonium from spent reactor fuel.[193] China remains the one NPT nuclear-weapon state not to formally espouse a moratorium on fissile material production. Kobayashi Yuki, a Japanese policy scholar who has worked in recent years on nuclear safeguards, observes, "[l]ack of transparency in plutonium production and nuclear arms expansion... may emaciate the IAEA's regime for international nuclear materials management."[194] Dr. Yuki no doubt is right to conclude that such thinning out of the IAEA's regime, in turn, would put the NPT as a whole at risk.[195] China's step away from fuel-cycle accountability endangers the NPT. However, it also presents an opportunity for future NPT negotiations. How might the United States pursue the opportunity?

One way is to continue negotiations toward a Fissile Material Cut-off Treaty (FMCT).[196] Having ceased to report on its plutonium

---

192. See "Concerns grow over China nuclear reactors shrouded in mystery," *Al Jazeera* (May 19, 2021): https://www.aljazeera.com/economy/2021/5/19/concerns-grow-over-china-nuclear-reactors-shrouded-in-mystery
193. Henry D. Sokolski, China's Civil Nuclear Sector: Plowshares to Swords? (2021) pp. 6-22.
194. Yuki Kobayashi, "Observations on Lack of Transparency in China's Nuclear Arms Expansion: Ahead of the NPT Review Conference," *SPF China Observer,* no. 39, August 17, 2022: https://www.spf.org/spf-china-observer/en/document-detail039.html#documentNote4
195. *Id.*
196. GA res. 48/75 of December, 1993, recommended the negotiation of a treaty to ban the production of fissile material for nuclear weapons. The effort to negotiate such a treaty was endorsed by the NPT Rev. Cons in 1995, 2000, 2005, 2010 and in GA res. 71/259 ("Treaty banning the production of fissile material for nuclear weapons or other nuclear explosive devices") of December 23, 2016.

holdings, and evidently increasing its plutonium extraction efforts, China has impeded steps toward an FMCT. However, China's conduct also offers a possible opening. For example, the Vienna Group of Ten, among other non-nuclear weapon states, express renewed support for the FMCT.[197] China itself expresses a "positive attitude" towards the FMCT.[198] So the United States is not proceeding on a blank slate when raising fissile-cutoff as an area meriting negotiation. The time seems right "to seek negotiations on a cut-off in the production of fissile material for use in nuclear weapons or other nuclear explosive devices."[199]

But a general call for FMCT negotiations will not lead to any practical result, if the prospective negotiating partners fail to settle key details. One key detail is identifying the proper framework in which to carry out FMCT negotiations. China insists that FMCT negotiations be confined to the UN Conference on Disarmament.[200] China's fixed position in this regard perhaps could be highlighted to illustrate the comparative flexibility of the United States. To do so, however, the United States would have to offer meaningful alternatives to China's position, for example a range of more flexible options as to the framework in which FMCT negotiations might take place.

The United States also would need to propose a substantive topic relevant to fissile cutoff that is not so intractable as to frustrate any

---

197. See NPT/CONF.2020/PC.III/WP.2 (March 15, 2017), Working paper submitted by Australia, Austria, Canada, Denmark, Finland, Hungary, Ireland, the Netherlands, New Zealand, Norway and Sweden (the Vienna Group of Ten), para. 7.
198. NPT/CONF.2020/PC.III/8 (April 29, 2019), Implementation of the Treaty on the Non-Proliferation of Nuclear Weapons of the People's Republic of China, Report submitted by China, para. 23 (referring to Shannon report (CD/1299)).
199. Ambassador Bonnie Denise Jenkins, Under Secretary for Arms Control and International Security, Remarks, Washington, DC, May 26, 2022: https://www.state.gov/priorities-regarding-the-new-and-emerging-challenges-to-international-security/
200. https://fissilematerials.org/blog/2020/05/no_clear_path_forward_for.html

eventual talks. Calling for greater accountability in the management of China's fuel cycle would be a modest, but meaningful, start. Calling for greater accountability would be reasonable, because even if China agreed merely to return to the reporting practice it had maintained prior to 2017, then progress would have been achieved against the impasse that China's retreat from accountability has caused. Negotiating on this point would not be asking China to do something China has not already done. Instead, it would note China's commitment in 1997 to the IAEA to report on transfers of plutonium to civilian use[201] and encourage China once more to observe that commitment.

Properly communicated, well-grounded concerns about regional security should give added impetus to NPT negotiations on China's plutonium production and lack of fuel-cycle accountability. Under the Rubio-Markey Act of 2016, it was proposed that Congress find that "China's plans to reprocess United States-origin spent fuel raise the risk that China could divert plutonium to military purposes, enabling it to produce additional nuclear weapons that threaten the United States and its allies" and that "China's pursuit of commercial plutonium reprocessing will increase the likelihood that Japan expands its commercial reprocessing program and that South Korea will increase efforts to initiate a similar program of its own."[202] The

---

201. China committed "to submit to safeguards by the IAEA on a voluntary basis… any plutonium transferred from military activities to peaceful nuclear activities": Information note of the People's Republic of China to the IAEA (Dec. 1, 1997): https://www.iaea.org/sites/default/files/infcirc549.pdf#page=3
202. S. 3010 – U.S.-China Nuclear Cooperation and Nonproliferation Act of 2016, introduced May 26, 2016, section 2, subparagraphs (9) and (11). See also, regarding Japan's Rakkasho plutonium recycling facility and its possible effects on fuel enrichment activities in China and South Korea, Henry Sokolski, *Underestimated: Our Not So Peaceful Nuclear Future* (Arlington: Nonproliferation Policy Education Center, 2015) 78-85. See also Victor Gilinsky & Henry Sokolski, "Make US-Japanese nuclear cooperation stable again: End reprocessing," *Bulletin of the Atomic Scientists* (June 27, 2018).

risk of regional escalation of this kind is one that China itself is best positioned to address. That risk is a further factor to which the United States should refer in making the case that China join negotiations on plutonium production and fuel-cycle accountability.

The 2022 NPR calls for China to "adopt a moratorium on fissile material production or, at a minimum, provide increased transparency to assure the international community that fissile material produced for civilian purposes is fully accounted for and not diverted to military uses."[203] However, as is the case with its calls for transparency in general in regard to China's nuclear weapons buildup, the 2022 NPR does not connect fuel cycle accountability to a renewed NPT negotiating agenda, or at least not in clear enough terms to contribute to the revitalization of the NPT that this paper argues it is in our strategic interests to pursue. Calling out China for its radical increase in fissile material production capacity, as the NPR does, is necessary, but in addition to unilateral statements that call China out the United States should be calling China to the table to negotiate.

(E) Safeguards for Fast Reactors and Keeping the IAEA in the Picture

IAEA safeguards under NPT Article III "are fundamental to the nuclear non-proliferation regime and help create an environment conducive to nuclear cooperation."[204] Article III formally addresses only the non-nuclear weapon states. The nuclear-weapon

203. 2022 NPR at 17.
204. Prep. Com. for the 2020 Rev. Con., 15 March 2017, NPT/CONF.2020/PC.III/WP.2, Addressing "Vienna issues", Working paper submitted by Australia, Austria, Canada, Denmark, Finland, Hungary, Ireland, the Netherlands, New Zealand, Norway and Sweden (the Vienna Group of Ten). Background Note 2, para. 1.

States long have stated publicly that they appreciate, however, that Article III's formal addressees—the non-nuclear weapon states—gain confidence in the IAEA safeguard system when they, too—the nuclear-weapon states—accept a responsibility to account for their nuclear activities. Even as IAEA safeguards are no panacea,[205] and safeguards have little effect unless *national* authorities implement them ably and in good faith, the continued vitality of the NPT requires the continued observance of NPT Article III.

China, in 2019 in one of its NPT Preparatory Committee reports, boasted that it has "actively reduced the use of sensitive nuclear materials and has completed the decommissioning of two domestic research microreactors and a low-enrichment retrofit project for a research microreactor."[206] China also mentioned its "demonstration fast-reactor project."[207] China did not say anything about IAEA access to that project. In the same Preparatory Committee report on NPT implementation, China claimed that "China adheres to the 'closed cycle' model of nuclear fuel development and has essentially completed its nuclear fuel cycle system, with the supply of nuclear fuel able to meet the fuel requirements of the nuclear power plants that have been put into operation."[208] This claim is, at best, selective in the account it gives of China's nuclear fuel activities. China does not acknowledge the real issue: China is massively expanding its capacity to produce fissile material.

---

205. A number of countries, not least of these Iraq and North Korea, have pursued covert nuclear weapons programs, notwithstanding IAEA safeguards. See Andrea Stricker, "International Atomic Energy Agency" in Richard Goldberg (ed) (with foreword by Amb. Nikki R. Haley), *A Better Blueprint for International Organizations. Advancing American Interests on the Global Stage* (FDD Press, Washington, DC: June 2021) 26.
206. NPT/CONF.2020/PC.III/8 (April 29, 2019), Implementation of the Treaty on the Non-Proliferation of Nuclear Weapons of the People's Republic of China, Report submitted by China, para. 63.
207. *Id.* at paras. 43, 87.
208. *Id.* at para. 87.

Negotiating for a more rigorous safeguarding of China's nuclear activities would be no mere cosmetic exercise. The United States understands that China very well may be reprocessing fuel from nuclear power plants and using civil reactors to produce tritium. Such reprocessing could violate China's peaceful end use agreements with the United States and arguably, too, would violate China's 1997 voluntary reporting understanding with the IAEA.[209] Short cycling and reprocessing of spent fuel from China's U.S. AP 1000 reactors and Westinghouse-designed Chinese variants has raised justified concern in the U.S. Senate.[210]

The need has grown for the nuclear-weapon states to demonstrate their commitment to the safeguards system, as the number of states has grown that either make enriched uranium, reprocess spent reactor fuel, or express that they are at liberty to do so.[211] This trend imposes costs on the IAEA. Fully 20 per cent of the IAEA budget in 2010 was allocated to monitoring the one plutonium reprocessing plant in a non-nuclear weapon state—Japan's reprocessing plant at Rokkash.[212] If the non-nuclear-weapon states perceive the nuclear-weapon states as indifferent to the safeguards system, then justifying the expenditures and maintaining transparency will become much more difficult. For these reasons, China's discontinuation of IAEA reporting and rapid buildup of plutonium processing capacity is further cause for concern.

---

209. NPEC Occasional Paper 2201 at p. 7.

210. *Id.*

211. Even seven years ago, a confident estimate was that Japan, South Korea, Argentina, Brazil, South Africa, Egypt, Turkey, Saudi Arabia, Iran, Vietnam, and Jordan numbered among States that either make enriched uranium, reprocess spent reactor fuel, or express that they are at liberty to do so. Henry Sokolski, *Underestimated: Our Not So Peaceful Nuclear Future* (Arlington: Nonproliferation Policy Education Center, 2015) 120.

212. See, UNIDIR, *A Fissile Material Cut-off Treaty. Understanding the Critical Issues* (2010) p. 33: https://unidir.org/sites/default/files/publication/pdfs//a-fissile-material-cut-off-treaty-understanding-the-critical-issues-139.pdf#page=48

The United States might explore a pause on commercial deployment of fast reactors and spent fuel recycling in the Pacific Rim, a natural step taking its lead from the Rubio-Markey 2016 proposed legislation.[213] The United States might also propose—or one or more non-nuclear-weapon States might propose—that all nuclear-weapon states open their civil nuclear facilities to IAEA inspection.[214] The United States long has been the leader among the NPT nuclear-weapon states when it comes to transparency. We should consider ways to leverage our strength in that area to revitalize the negotiating agenda under NPT Article VI.

We also should continue to implement measures to ensure that proposed advanced modular fast reactors and recycling plants operate under safeguard agreements. Nonproliferation officials during the Trump Administration pursued a number of helpful steps toward safeguards for new classes of civil nuclear plant. For example, under the FIRST program (Foundational Infrastructure for Responsible Use of Small Modular Reactor Technology), which the Biden Administration has continued, nonproliferation officials engage with foreign partners to lay the groundwork for nonproliferation-compliant civil nuclear energy projects.[215] The Trump Administration and Biden Administration alike also have concluded a number of Nuclear Cooperation Memoranda of Understanding (NCMOUs), non-binding bilateral agreements that

---

213. S. 3010 – U.S.-China Nuclear Cooperation and Nonproliferation Act of 2016, introduced May 26, 2016, section 9(a) stating that "it is the sense of Congress that the United States should encourage countries in East Asia to forego the commencement of new spent fuel reprocessing activities as part of a mutual effort to prevent the increased or expanded stockpiling of separated plutonium in the region." Section 9(a).
214. A proposal aired in NPEC Occasional Paper, p. 12.
215. See, e.g., U.S.-Latvia Joint Statement on the New Clean Energy and Nuclear Security Collaboration (Apr. 4, 2022): https://www.state.gov/joint-statement-on-the-new-clean-energy-and-nuclear-security-collaboration-under-the-foundational-infrastructure-for-responsible-use-of-small-modular-reactor-technology-first-initiative/

set the stage for civil nuclear energy with foreign partners in a framework that implements safety, security, and nonproliferation standards.[216] Programs such as these demonstrate the United States's commitment to the Article III guarantee of access to peaceful nuclear energy. They also demonstrate that we take our Article I nonproliferation obligations seriously. A renewed negotiations agenda under Article VI could include discussions toward ensuring that China does no less.

(F) Nuclear Testing—Let's Talk Definitions (CTBT Framework)

As the Congressional Research Service recalled some years back, "[a] ban on nuclear testing [is] the oldest item on the [nuclear] arms control agenda."[217] Negotiations toward a ban on nuclear testing have been closely tied to the NPT from the start. In particular, as the time approached to decide whether to extend the NPT indefinitely,[218] NPT non-nuclear weapon states expressed dissatisfaction with the absence of a test ban, and they asserted that the failure to adopt one underscored a fundamental unfairness in the Treaty. NPT non-nuclear weapon states said that they had given up a great deal by supporting the NPT—what many of them said otherwise would have been a sovereign right to acquire nuclear weapons; but (they said) the nuclear-weapon states had

---

216. See Fact Sheet: https://www.state.gov/nuclear-cooperation-memoranda-of-understanding-ncmou/
217. Congressional Research Service, *Comprehensive Nuclear-Test-Ban Treaty: Background and Current Developments* (September 1, 2016) p. 1. Cf. *id.* at Summary.
218. NPT Article X(2) provided that "[t]wenty-five years after the entry into force of the Treaty, a conference shall be convened to decide whether the Treaty shall continue in force indefinitely, or shall be extended for an additional fixed period or periods. This decision shall be taken by a majority of the Parties to the Treaty." 729 U.N.T.S. at p. 175.

given up nothing. The non-nuclear weapon states argued, in short, that the time had come to remedy an imbalance in the NPT bargain. A future ban on nuclear tests, in the eyes of the NPT non-nuclear weapon states, came to be "the touchstone of good faith on these matters."[219]

At the NPT Review and Extension Conference of April-May 1995, non-nuclear weapon states insisted that the nuclear-weapon states make a material commitment to offset the perceived unfairness in the NPT bargain. As a result, the Conference elevated the achievement of a test ban to a matter of priority. The Final Document of the 1995 Review and Extension Conference connected the achievement of a test ban treaty expressly to NPT Article VI. According to the Final Document,

> The achievement of the following measures is important in the full realization and effective implementation of article VI…
>
> (a) The completion by the Conference on Disarmament of the negotiations on a universal and internationally and effectively verifiable Comprehensive Nuclear-Test-Ban Treaty no later than 1996…[220]

Thus, on its express terms, the Final Document indicated that achieving a test ban is "important" if Article VI is to be fully realized and effectively implemented. The Final Document further described the prospective test ban as helping the states parties "fulfill nuclear disarmament "with determination."[221]

---

219. CRS (2016) at p. 22.
220. NPT/CONF.1995/32 (Part I) at p. 10.
221. *Id.* at p. 9.

Negotiations toward a nuclear test ban treaty[222] took place in the frame of the UN Conference on Disarmament and resulted in a draft Comprehensive Nuclear-Test-Ban Treaty (CTBT),[223] which Australia duly submitted to the UN General Assembly.[224] The UN General Assembly adopted the CTBT on September 17, 1996 and called on "all States to sign and... become parties to" it.[225] As can be seen from Australia's engagement on the matter in the General Assembly (and also from the support for the drafting process in the Conference on Disarmament by other U.S. allies),[226] countries friendly to the United States understood the CTBT to be an important step. Indeed, some twenty years later, as the 2020 NPT Review Conference approached, key U.S. partners and allies continued to lay emphasis on the CTBT, describing it, for example, as an "integral part of the 1995 decision to indefinitely extend" the NPT and "a core element of the nuclear disarmament and nuclear non-proliferation regime."[227] At the time of its adoption, non-

---

222. As to the negotiations toward the CTBT, see Jaap Ramaker, Jenifer Mackby, Peter Marshall & Robert Geil, *The Final Test: A History of the Comprehensive Nuclear-Test-Ban Treaty Negotiations* (Vienna: Provisional Technical Secretariat of the Preparatory Commission for the Comprehensive Nuclear-Test-Ban Treaty Organization, 2003).
223. Conference on Disarmament doc. CD/1427 (August 22, 1996): https://documents-dds-ny.un.org/doc/UNDOC/GEN/G96/636/59/img/G9663659.pdf?OpenElement
224. A/50/1027 (August 26, 1996): https://documents-dds-ny.un.org/doc/UNDOC/GEN/N96/219/23/img/N9621923.pdf?OpenElement
225. A/RES/50/245, September 1996, paras. 1 & 3.
226. See CD/1429, op cit.
227. NPT/CONF.2020/PC.III/WP.2 (March 15, 2017), Addressing "Vienna issues", Working paper submitted by Australia, Austria, Canada, Denmark, Finland, Hungary, Ireland, the Netherlands, New Zealand, Norway and Sweden (the Vienna Group of Ten). See also Non-Proliferation and Disarmament Initiative States Working paper, reaffirming on behalf of "a diverse cross-regional grouping of non-nuclear weapon States" their "strong commitment to strengthening the nuclear test ban regime" and their "strong concern that the [CTBT] still has not entered into force after 20 years": NPT/CONF.2020/PC.I/WP.3 (March 17, 2017).

aligned countries also expressed their support,[228] which they have continued.[229]

China, at the time of the CTBT's adoption in 1996, while supporting the CTBT, said that the CTBT "is not entirely satisfactory."[230] Tellingly, China sought to justify its misgivings by referring to what the CTBT does *not* do: "it does not touch upon the conclusion of a convention on the comprehensive prohibition of nuclear weapons."[231] China since has joined the other nuclear-weapon states in rejecting the TPNW, but a comprehensive prohibition against nuclear weapons, now pursued through the TPNW, historically found support from the NPT nuclear-weapon

---

228. For example, Tanzania, a long-standing stalwart of the non-aligned movement, though abstaining from the General Assembly vote on adoption of the CTBT because (according to Tanzania) "[t]he treaty perpetuates the status quo by allowing the most technologically sophisticated nuclear-weapon States to continue with the vertical proliferation of nuclear arsenals through computer simulation" and because it "provides for [no] further negotiations," nevertheless voiced its endorsement of a test-ban as part of incremental progress toward eventual disarmament:

> "Tanzania has all along been a staunch supporter and advocate of a comprehensive test-ban treaty (CTBT). We believed in the CTBT, and indeed we have always considered that it was *the only viable first step* leading to the total elimination of nuclear weapons" (emphasis added). Mr. Mwakawago (United Republic of Tanzania): A/50/PV.125 (September 10, 1996) at p. 1.

As to Tanzania's prominence in the non-aligned movement, see T.V. Sathyamurthy, "Tanzania's Non-Aligned Role in International Relations," 37(1) *India Quarterly* 1-23 (1981).

Pakistan, to give another example, "consistently supported the objective of a comprehensive nuclear-test ban *as an essential step towards* nuclear disarmament and as a means of promoting nuclear non-proliferation." Mr. Akram (Pakistan), *id.* at p. 7.

229. See, e.g., Final Document of the 18th Non-Aligned Movement Summit of Heads of State and Government, Baku (October 25-26, 2019), NAM 2019/CoB/Doc.1, p. 81 para. 244. For an earlier Non-Aligned Movement affirmation, see CTBT Organization Preparatory Commission, "Non-aligned Movement Conference Stresses Importance of Comprehensive Nuclear-Test-Ban Treaty, PI/2003/02 (2003).

230. Mr. Sha Zukang (China), A/50/PV.125 (September 10, 1996) at p. 10.

231. *Id.*

states *least* likely to be constrained by the suasive effects of such a prohibition. As this paper noted above, the prohibitionist agenda has asymmetric effects between states that have open political systems and those that do not.

The United States signed the CTBT on September 24, 1996.[232] The United States Senate has not given its consent to ratification. China and the United States are the two NPT nuclear-weapon states not to have ratified the CTBT (though other states, too, among the forty-four whose ratifications are needed before the CTBT enters into force have likewise refrained from ratifying).[233] The continued non-observance of its CTBT commitments by Russia and the high level of nuclear activity by China at its Lop Nur nuclear weapons test site in recent years have validated the Senate's hesitancy to consent to ratification.[234]

Yet support for the CTBT continues, with partners and allies, as well as non-aligned countries, lamenting that the CTBT still has not entered into force. For example, in the Preparatory Committee for the 2020 NPT Review Conference, countries comprising the "Non-Proliferation and Disarmament Initiative" (Australia, Canada, Chile, Germany, Japan, Mexico, the Netherlands, Nigeria,

---

232. For a list of States having signed, including those having ratified, the CTBT, see
https://treaties.un.org/Pages/showDetails.aspx?objid=0800000280049f7f&clang=_en
233. Under CTBT Article XIV, ratification by the States listed in Annex 2 to the Treaty (A/50/1027 at p. 48) is required before the CTBT enters into force. Annex 2 lists forty-four States, among them all five NPT nuclear-weapon States. Among the Annex 2 States, Egypt, Iran, and Israel (along with China and the United States) have signed but not ratified the CTBT; North Korea, India, and Pakistan have not signed.https://www.nti.org/education-center/treaties-and-regimes/comprehensive-nuclear-test-ban-treaty-ctbt/#:~:text=After%20this%20ratification%2C%20nine%20Annex,Pakistan%2C%20and%20the%20United%20States.
234. See U.S. Department of State, *Executive Summary of Findings on Adherence to and Compliance with Arms Control, Nonproliferation, and Disarmament Agreements and Commitments* (April 2020), p. 8.

the Philippines, Poland, Turkey and the United Arab Emirates) made known their "strong concern" that over twenty years had elapsed since its adoption without the CTBT entering into force.[235]

Efforts to advance the substance of the CTBT would demonstrate that the United States recognizes the concern that these countries and others have expressed. A crucial element of substance is the definition of "nuclear test."

Significant obstruction and delay in arms control have resulted from contested definitions. Russia skirmishes over definitions as a way to evade its arms control obligations. The most conspicuous example are strategic offensive systems that Russia has newly developed but insists "have nothing to do with the strategic offensive arms categories covered by the [New START] Treaty."[236] The 2022 NPR observes that "Russia is pursuing several novel nuclear-capable systems designed to hold the U.S. homeland or Allies and partners at risk."[237] These systems include Russia's Burevestnik nuclear-powered cruise missile, Kinzhal air-launched ballistic missile, and Status-6 Poseidon nuclear torpedo. Then-U.S. Under Secretary for Arms Control and International Security Andrea Thompson stated to Congress in 2019 that the United States considers these three systems to be "new kinds of strategic offensive arms."[238] Russia says "we find it inappropriate

---

235. NPT/CONF.2020/PC.I/WP.3 (March 17, 2017), Comprehensive Nuclear-Test-Ban Treaty, Working paper submitted by the members of the Non-Proliferation and Disarmament Initiative (Australia, Canada, Chile, Germany, Japan, Mexico, the Netherlands, Nigeria, the Philippines, Poland, Turkey and the United Arab Emirates).

236. https://www.wsj.com/public/resources/documents/Russia.pdf?mod=article_inline#page=9

237. 2022 NPR at 4 (emphasis added).

238.https://www.foreign.senate.gov/imo/media/doc/05%2015%2019%20Intermediate-Range%20Nuclear%20Forces%20Treaty.pdf#page=47 Hearing before the Committee on Foreign Relations, United States Senate, 116th Cong. 1st Sess. (May 15, 2019) at page 43.

to characterize new weapons being develop[ed] by Russia that do not use ballistic trajectories of flight moving to a target as 'potential new kinds of Russian strategic offensive arms.'"[239] So definitions matter.

China touts its role in the P5 Working Group on the Glossary of Key Nuclear Terms.[240] China also continues to profess support for "the purposes and objectives" of the CTBT.[241] Yet significant ground remains to be covered, before working definitions relevant to nuclear weapons tests are agreed.[242] Without agreement as to definitions, all the technical challenges of verification and compliance in regard to a test moratorium are compounded. Semantic differences will impede implementation, even if all negotiating parties proceed in perfect good faith. Bringing definitions concerning nuclear tests into the NPT negotiating agenda would be a meaningful step, in line with the long-standing support for the CTBT among U.S. allies, partners, and other countries.

239.https://www.wsj.com/public/resources/documents/Russia.pdf?mod=article_inline#page=9

240. See, e.g., Implementation of the Treaty on the Non-Proliferation of Nuclear Weapons in the People's Republic of China, Report submitted by China: NPT/CONF.2020/PC.III/8 (April 29, 2019) para. 35. For the glossary, see *P5 Glossary of Key Nuclear Terms* (Beijing: China Atomic Energy Press, April 2015): https://2009-2017.state.gov/documents/organization/243293.pdf

241. NPT/CONF.2020/PC.III/8, para. 18.

242. Kathleen Bailey & Thomas Scheber, *The Comprehensive Test Ban Treaty: An Assessment of the Benefits, Costs, and Risks* (Fairfax, Virginia: National Institute for Public Policy, March 2011) pp. 15-16; Mark B. Schneider, "Yes, the Russians are Testing Nuclear Weapons and It Is Very Important," *RealClearDefense* (August 8, 2019): https://www.realcleardefense.com/articles/2019/08/08/yes_the_russians_are_testing_nuclear_weapons_and_it_is_very_important_114651.html

(G) Nuclear Safety

With China's civil nuclear generating output fast approaching that of France, the country long holding the record,[243] specialists have raised questions about the safety and governance of nuclear power in China. Philip Andrews-Speed, for example, says that areas of concern in China's civil nuclear industry include lack of capacity and of independence in China's National Nuclear Safety Administration; an incoherent legal framework for nuclear safety and security; deficient provision for civil nuclear liability, including in regard to nuclear accidents having international impact; and inadequate public consultations when it comes to the planning of new power plants.[244]

NPT parties have addressed nuclear safety in the civil sector under the NPT Article III rubric. The "Vienna Group of Ten," for example, in connection with the 2020 Review Conference addressed nuclear security, including "environmental safety and notification of coastal states of intended transit of nuclear materials."[245] Similar issues might merit inclusion as well in a revitalized negotiations agenda under Article VI. At least in a general way, these issues touch on the serious concerns that have emerged as evidence grows that China is diverting nuclear fuel

---

243. The World Nuclear Association, as of October 2022, reports that China's civil nuclear generating capacity is 52,150 megawatts. https://world-nuclear.org/information-library/country-profiles/countries-a-f/china-nuclear-power.aspx The Association reports that France's capacity is 61,370 megawatts: https://world-nuclear.org/information-library/country-profiles/countries-a-f/france.aspx

244. Philip Andrews-Speed, "The governance of nuclear power in China," 13(1) *The Journal of World Energy Law & Business* 23-46 (2020).

245. Prep. Com. for the 2020 Rev. Con., 15 March 2017, NPT/CONF.2020/PC.III/WP.2, Addressing "Vienna issues", Working paper submitted by Australia, Austria, Canada, Denmark, Finland, Hungary, Ireland, the Netherlands, New Zealand, Norway and Sweden (the Vienna Group of Ten), Background note 5: nuclear safety. Including environmental safety and notification of coastal states of intended transit of nuclear materials.

from its civil reactor program to weapons use.[246] To be sure, nuclear safety relies chiefly on national regulatory régimes and enforcement. Nevertheless, safety failures in the nuclear field are an international concern, and sound safety practices foster nonproliferation objectives as well. The United States and other negotiating partners thus may credibly refer to our concerns over China's safety rules and practices in the frame of Article VI.

(H) Export Controls

Experts have warned that China, in the pursuit of export markets in emerging technologies, ignores distinctions between civil and military end-use, thus frustrating one of the main policy objectives that motivates U.S. export controls.[247] As for illicit acquisition of nuclear materials and equipment for its own use, China's misconduct is well-established.[248] Evasion or avoidance of export controls is a serious challenge where *any* military application is concerned, but the challenge is all the more urgent, where nuclear materials and equipment are concerned.

---

246. As to fuel diversion, see Christopher Ford & Thomas Countryman, "Preface" to Henry. D. Sokolski, ed., *China's Civil Nuclear Sector: Plowshares to Swords?* (NPEC Occasional Paper 2102, March 2021) pp. 1-2: http://npolicy.org/article_file/2102_Chinas_Civil_Nuclear_Sector.pdf; Frank von Hippel, "Projecting Plutonium Stocks to 2040," Chapter 4 in Sokolski, ed. (March 2021) at pp. 65-86. For a sceptical assessment, see Hans M. Kristensen & Matt Korda, "Chinese nuclear weapons, 2021," 77(6) *Bulletin of the Atomic Scientists* 318-319 (2021).

247. See Congressional Research Service, *U.S. Export Controls and China* (updated March 24, 2022): https://crsreports.congress.gov/product/pdf/IF/IF11627

248. See, e.g., "DOE Announces Measures to Prevent China's Illegal Diversion of U.S. Civil Nuclear Technology for Military or Other Unauthorized Purposes," U.S. Department of Energy, October 11, 2018: https://www.energy.gov/articles/doe-announces-measures-prevent-china-s-illegal-diversion-us-civil-nuclear-technology

A number of Allies and like-minded countries have identified nuclear export controls as an area meriting attention in the frame of the NPT. For example, in a joint paper submitted during the Preparatory Committee meetings leading up to the 2020 NPT Review Conference, the Vienna Group of Ten[249] stated, "Nuclear export controls are a legitimate, necessary and desirable means of implementing the obligations of States parties under Article III."[250] The troublesome admixture by China of civil programs and military applications, including through its Military-Civil Fusion doctrine,[251] further suggests that good faith negotiations on export controls in the nuclear field would be a fruitful step for arms control and nonproliferation.

---

249. Australia, Austria, Canada, Denmark, Finland, Hungary, Ireland, the Netherlands, New Zealand, Norway, and Sweden. The Vienna Group of Ten has convened for NPT Prep.Com. and Rev. Con. meetings since 1980 to advance agreed nonproliferation issues that they refer to as "Vienna issues." See Vienna Group of Ten, May 2017 NPT Prep.Com., General Debate Statement: https://geneva.mission.gov.au/gene/Statement776.html

250. Prep. Com. for the 2020 Rev. Con., 15 March 2017, NPT/CONF.2020/PC.III/WP.2, Addressing "Vienna issues", Working paper submitted by Australia, Austria, Canada, Denmark, Finland, Hungary, Ireland, the Netherlands, New Zealand, Norway and Sweden (the Vienna Group of Ten), Background note 3: export controls, para. 1.

251. See Christopher A. Ford, *The PRC's "Military-Civil Fusion" Strategy Is a Global Security Threat,* U.S. Department of State Dipnote (March 16, 2020) (delivered in Dr. Ford's capacity as Assistant Secretary for International Security and Nonproliferation and performing the duties of the Under Secretary of State for Arms Control and International Security, https://2017-2021.state.gov/the-prcs-military-civil-fusion-strategy-is-a-global-security-threat/index.html Concern over China's Military-Civil Fusion is not confined to U.S. government circles. See, e.g., Manoj Joshi, "China's Military-Civil Fusion Strategy, the US Response, and Implications for India," *Observer Research Foundation Occasional Papers* (January 25, 2022), https://www.orfonline.org/research/chinas-military-civil-fusion-strategy/. The concerns over Military-Civil Fusion extend beyond traditional defense considerations, affecting as that doctrine does a range of U.S. interests, such as patent protection: see Thomas D. Grant & F. Scott Kieff, "Great Powers and New Risks: What Businesses and Regulators Should Know about China's Strategic Ambitions," 65(2) *Orbis* 257, 262-270 (February 2021).

(I) Institutionalizing the NPT

The Chemical Weapons Convention, the Biological Weapons Convention, and the Comprehensive Test Ban Treaty each has its own permanent organization staffed and funded to provide a focal point for diplomatic engagement and, where needed, for seeking solutions in the face of credible evidence of non-compliance.[252] The NPT, though kindred to these other instruments, concerned as all of them are with weapons of mass destruction, has no such institutional expression. Henry Sokolski, in a recent retrospective on the NPT in the American Bar Association's National Security Law forum, suggests that a permanent NPT administrative body would assist in the implementation of the Treaty.[253] In considering potential topics for negotiation—and potential steps to revive the NPT—the United States could explore the creation of such an institution.

---

252. As to the Chemical Weapons Convention, see https://www.opcw.org/about/mission ; the Preparatory Commission to the Comprehensive Test Ban Treaty Organization, https://www.ctbto.org/our-mission ; and the Biological Weapons Convention Implementation Support Unit, https://www.un.org/disarmament/biological-weapons/implementation-support-unit/relevant-activities-overseen-by-the-isu

253. See Henry D. Sokolski, "Nuclear Non-Proliferation… If You Can Keep It" (Jan. 6, 2023), *ABA Standing Committee on Law & National Security*: https://www.americanbar.org/groups/law_national_security/publications/aba-standing-committee-on-law-and-national-security-60-th-anniversary-an-anthology/nuclear-non-proliferation-if-you-can-keep-it/

(J) The Conventional-nuclear Deterrence Matrix: Strengthening the Whole by Moderating Reliance on the Nuclear Quantitative Part

China says that nuclear-weapon States should "diminish the role played by nuclear weapons in their national security policies."[254] In drawing attention to the conventional-nuclear deterrence matrix, China opens the door to a potentially meaningful line of negotiations: China's increasing threats of conventional armed force in its region, not least of all against Taiwan, merit discussion. The United States should entertain raising the matter in the NPT Article VI frame.

Pertinent here, success in negotiating nuclear issues could enhance security overall. For instance, negotiating fissile material accountability with China would demonstrate commitment to NPT Article III which, in turn, would add credibility when the United States and our allies call on others, such as Saudi Arabia, to give responsible accounts of their own civil nuclear projects.[255] Enhancing conventional deterrence for the benefit of countries that otherwise entertain the pursuit of nuclear weapons might well have synergy, in turn, with reinvigorated efforts under Article III.[256]

---

254. NPT/CONF.2020/PC.III/WP.36 (April 26, 2019), Security assurance, Working paper submitted by China, para. 3.
255. See Henry Sokolski, "Will Seoul and Washington make Riyadh nuclear-weapons ready?" *Bulletin of the Atomic Scientists* (July 26, 2022) (addressing, *inter alia,* concerns over the 2011 South Korean-Saudi Arabia nuclear cooperative agreement authorizing Saudi Arabia to enrich uranium up to 20 per cent without Korean consent): https://thebulletin.org/2022/07/will-seoul-and-washington-make-riyadh-nuclear-weapons-ready/ For the 2011 agreement, which entered into force August 14, 2012, see 3152 U.N.T.S. 14: https://treaties.un.org/doc/Publication/UNTS/No%20Volume/54043/Part/I-54043-080000028049247d.pdf#page=14
256. This paper has noted already, above, the benefits of persuading U.S. allies in the Pacific, in particular South Korea and Japan, to refrain from the possible future pursuit of nuclear weapons. The recognition by other U.S. allies in the

Quantitative nuclear arms-racing is expensive. A shift of the deterrence matrix toward nuclear weapons quantity will divert fiscal appropriations from other efforts, including advanced conventional weapons. The net effect of such diversion almost certainly would be detrimental to U.S. security: the battlefield evidence from Ukraine and elsewhere confirms that U.S. investments in advanced conventional weapons have been well-placed (even if by no means adequate for all contingencies). The NPT Article VI frame can include negotiations toward possible arms limitations (subject of course to the essential need for reliable verification). If such negotiations succeeded, then we would have an opportunity to moderate our reliance on the nuclear quantitative dimension of deterrence. That China itself asserts that it would like nuclear weapons to play a diminished role in security provides an opening. The United States should consider using NPT Article VI negotiations to put China to the test in regard to the nuclear quantitative dimension.

VI. Conclusion: An NPT Future and Bringing Realists Back to Arms Control

The task of revitalizing the NPT must be seen in light of the danger that its fading from the scene presents. Without the NPT, a vacuum will result that the TPNW will fill. Avoiding that state of affairs will not be easy. As this paper has suggested, disarmament activists and nuclear security "hawks" partake, by turns, in indifference and antipathy toward the NPT. The NPT has

---

region that investment in conventional forces benefits their security is helpful in this regard. See Mick Ryan, "The pros and cons of hosting B-52 on our shores," *The Sydney Morning Herald* (November 1, 2022). Ryan, a retired major general in the Australian Defence Force, argues that basing U.S. B-52s in Australia's Northern Territory allows Australia to appropriate funds to conventional weapons which he identifies as "more cost-effective ways to construct a strategic deterrent."

disappointed the activists, because it has not led to nuclear disarmament. The NPT either holds no interest for defense-minded foreign policy thinkers, because its Article VI negotiating clause seems too abstract to mean very much, or it arouses their mistrust, because, though it originated in realists' insight about how nuclear deterrence works, it long served as rally point for the disarmament community. This is why the NPT today is indeed nearly an orphan on the American foreign policy and defense landscape.

And, yet, though this paper has drawn attention to the near orphaning of the NPT, traces of support for the NPT still exist in divergent corners of American foreign policy. As the author of the present paper has written (with Henry Sokolski),[257] the Biden Administration evidently agrees with the Trump Administration that China's refusal to join nuclear arms control talks is highly problematic, arguably a breach of NPT Article VI.[258] As noteworthy as is the Biden Administration's public affirmation that China threatens the stability that long had prevailed among nuclear-weapon states, noteworthy too is the readiness of staunch exponents of nuclear defense to invoke the NPT. Marshall Billingslea, who served as Special Presidential Envoy for Arms Control in the Trump Administration, declared that "China is perilously close to standing in direct violation of the NPT."[259] Together with his then-principal, Secretary of State Mike Pompeo,

---

257. Henry Sokolski & Thomas D. Grant, "China: a nuclear nonproliferation deadbeat," *The Hill* (August 30, 2022): https://thehill.com/opinion/national-security/3621641-china-a-nuclear-nonproliferation-deadbeat/
258. President Biden, on the occasion of the 10th NPT Rev. Con., called on China to observe its obligations under the NPT: President Biden Statement Ahead of the 10th Review Conference of the Treaty on the Non-Proliferation of Nuclear Weapons (August 1, 2022): https://www.whitehouse.gov/briefing-room/statements-releases/2022/08/01/president-biden-statement-ahead-of-the-10th-review-conference-of-the-treaty-on-the-non-proliferation-of-nuclear-weapons/
259. Amb. Marshall S. Billingslea, "Arms Control and the New Start Treaty" (December 8, 2020): https://nipp.org/information_series/billingslea-marshall-arms-control-and-the-new-start-treaty-information-series-no-472/

Ambassador Billingslea drew further attention to Article VI in particular when calling out China for its nuclear buildup.[260] Senator Ted Cruz, among the staunchest exponents of American nuclear defense, meanwhile, is largely in concurrence with ICAN, the International Campaign to Abolish Nuclear Weapons, on the importance of negotiations under Article VI.[261] People and institutions who approach arms control from very different political starting points are arriving at the same conclusion; that they converge the way they do suggests the merits of placing renewed emphasis on the NPT. It also suggests that there is hope for bringing about an NPT revival. Despite widespread estrangement, policy makers have not yet disowned the NPT entirely.

Nevertheless, the disarmament community has come to occupy most of the space where NPT Article VI is concerned, and too many in the defense policy community have ignored Article VI altogether. Ignoring Article VI has been a high-risk gamble. The evidence is now accumulating that advocates of effective deterrence have lost the bet. If Article VI completes its slide toward desuetude, it will be the TPNW that fills the resultant vacuum. Ignoring Article VI, a crucial part of the American defense policy establishment will have ceded the nuclear space to the TPNW. This would be no mere tactical retreat. If the TPNW is all that we have left, then the nuclear deterrent will have lost its argumentative foundation. This would be a setback for American strategy and world security.

---

260. Michael R. Pompeo & Marshall Billingslea, "China's Nuclear Build-Up Should Worry the West," *Newsweek* (January 4, 2021): https://www.newsweek.com/chinas-nuclear-madness-opinion-1558342
261. See Senator Ted Cruz, "We should enact sanctions on China's defense industries unless they comply with their treaty obligations to halt their buildup," Twitter (May 18, 2021): https://twitter.com/sentedcruz/status/1394752627368513538; ICAN, "China's nuclear buildup violates international law," *ICAN* (November 12, 2021), https://www.icanw.org/china_nuclear_buildup

Thus, we need to make a prudential case for NPT Article VI. Is Article VI a perfect treaty provision? Far from it. But what happens if it recedes from view? The TPNW occupies the entire field. The TPNW is a text that states a *present* prohibition in non-negotiable terms. And the TPNW is no mere hypothetical future focal point. It is a treaty-in-force, and it continues to gain support among key constituencies.

A determined effort to bring new life to the NPT is called for. The traces of support that we hear across the political spectrum give reason to hope that the United States might well lead that effort. However, so far, we hear few, if any prescriptions from either side of the aisle as to specific, practical steps. This paper has argued that the negotiations clause of NPT Article VI is an obvious focal point for practical steps and a renewed security diplomacy.

We need to use the NPT proactively, and to do that, we need to achieve the widest possible buy-in across the spectrum of mainstream opinion in our political system. Using the NPT to bring pressure to bear on China for its largely unaccountable nuclear weapons buildup is an important step. We should continue using the NPT in that way. However, it is also time to revitalize our approach to negotiating under the NPT. To do *that,* we need to take a fresh look at the substance that we are ready to bring to the negotiating table. It is the very heart of the realists' insight that arms control and eventual disarmament are an endpoint attainable only if we first address the substantive differences that divide us from our nuclear rivals. This paper has suggested a number of issues over which substantive differences exist but which are sufficiently concrete and practical that they could lend themselves to solutions in good faith negotiation. Here, too, a fresh approach to the NPT can bring the realists back to arms control. The stakes today are too high for us not to join the endeavor.

Xi and Putin Are Building More Nukes: How to Compete

Henry Sokolski

Three years after *The Washington Post* first released photos of China's construction of more than 300 nuclear missile silos and the Pentagon forecasted China would have at least 1,000 nuclear weapons by 2030, our government is still pondering what it should do. Meanwhile, Russia is developing entirely new nuclear weapons systems — hypersonic nuclear-armed missiles, nuclear anti-satellite weapons, and long-range nuclear torpedoes. The Pentagon is modernizing America's nuclear bombers, submarines, and missiles. But that was decided as part of the approval of New START back in 2010.

It also won't be enough. Matching China and Russia's nuclear expansion goals missile-for-nuclear-missile, warhead-for-nuclear-warhead, much less clearly staying ahead, could require a tripling or more of America's current nuclear deployments of 1,770 nuclear weapons.

That's a tall order. America's nuclear weapons pit production, nuclear submarine, and intercontinental ballistic missile modernization programs have repeatedly run over budget and continue to slip behind schedule.[262] China, in contrast, has a proven track record of deploying new weapons systems at record pace. As for Russia, it has the world's largest nuclear weapons materials production base.

---

262. Stephen Young, "Another reason to cancel the Sentinel missile: the rising cost of its nuclear warhead," *Bulletin of the Atomic Scientists*, May 23, 2024. https://thebulletin.org/2024/05/another-reason-to-cancel-the-sentinel-missile-the-rising-cost-of-its-nuclear-warhead/.

Our closest allies know this. South Korea, Japan, and Poland want the United States to base nuclear weapons on their soil or to get nuclear weapons of their own. This presents two problems. First, overseas U.S.-based nuclear weapons sites are target sinks. The United States placed thousands of its weapons in Taiwan, Japan, South Korea, and NATO states in the 1950s and 1960s because the Pentagon lacked sufficient long-range missiles to defend its allies. Now, it does. That's why the Pentagon removed nearly all of its forward deployed nuclear weapons in Europe and Asia. Redeploying them would be a military step backwards.

Second, letting our closest allies go nuclear could increase alliance tensions. Historically, this happened when the British, French, Israelis, and Pakistanis went nuclear.[263] Yet, as China and Russia increase their nuclear capabilities, the demand signal for our allies to go nuclear is rising.

What, then, should we do?

Adjust America's Nuclear Strategy

America's military has long required its strategic nuclear forces to be able to deliver a devastating counter punch against Russia or China as soon as either launch a nuclear attack against the United States. The aim of such a counterstrike would be to limit additional damage to U.S. targets that Russia or China might otherwise inflict. This puts a premium on launching U.S. nuclear forces as soon as possible after a nuclear attack — either right after a warning of launch or immediately after Chinese or Russian nuclear weapons detonate on US soil.

These demanding requirements determined much of the quality and quantity of America's current nuclear force. Fifty years ago,

---

263. Zachary Keck, *Atomic Friends: How America Deals with Nuclear-Armed Allies*, Rowman and Littlefield, August 2022. https://npolicy.org/atomic-friends-how-america-deals-with-nuclear-armed-allies/.

the United States had over 31,000 nuclear weapons that were relatively immune from being knocked out. The US military aimed these weapons at a limited number of relatively soft, fixed Russian long-range nuclear systems, which made prioritizing damage limitation sensible.

Accomplishing this now is far more challenging. Many Chinese nuclear missiles are mobile and based in three thousand miles of deep underground tunnels. China is installing many more in hundreds of silos. China is also developing new, stealthy long-range bombers, advanced nuclear ballistic missile submarines, and hypersonic missiles.

Then there's China's ally – Russia. It has thousands of nuclear weapons mounted on mobile intercontinental ballistic missiles, strategic bombers, nuclear submarines, and theater nuclear missiles. Because China's and Russia's nuclear forces are large, relatively invulnerable, and growing, continuing to prioritize the mission of limiting damage could easily require doubling or tripling America's current deployed nuclear force of roughly 1,770 weapons.

That would be difficult. First, the Pentagon and Energy Department have struggled to update the current U.S. force of intercontinental ballistic missiles, nuclear warheads, ballistic missile submarines, and bombers. All of these programs are significantly behind schedule or over budget. The prospects of the United States quickly outproducing China and being confident in their ability to destroy remaining Russian and Chinese nuclear systems after a first strike are slim and dwindling. In the 1960s we had a clear shot; now it's unlikely.

Worse, the most important prompt strike missiles designed to achieve this damage limitation mission — America's intercontinental ballistic missiles — are themselves vulnerable as they are based in relatively soft, fixed silos. How many would survive a first strike is unclear unless they were launched on warning or under attack. As the recent Congressional Commission on the Strategic Posture of the United States noted, however, launching under these circumstances is hardly desirable.

This suggests adjusting America's nuclear strategy. The Pentagon should still try to limit damage after the United States is attacked and continue updating its current aging nuclear force. But it should also recognize that prioritizing damage limitation will become more expensive and difficult as the numbers of Russian and Chinese nuclear systems, their location uncertainty, and their known positions increase.

Rather than aiming to significantly expand the U.S. nuclear arsenal to achieve a high level of damage limitation, the United States should place higher priority on making it more difficult for China or Russia to disable or destroy U.S. and allied nuclear forces. This recommends transitioning all its siloed intercontinental ballistic missiles to mobile platforms and camouflaging and decoying them along with U.S. strategic submarines and bombers. Alternative basing on commercial planes, trains, and ships also should be explored. For key fixed targets, active defenses and hardening may be necessary.

These efforts aim to exponentially increase the number of possible aim points China and Russia would have to target to knock out America's strategic nuclear forces. Instead of Russia and China needing hundreds of nuclear-armed missiles and thousands of nuclear warheads to accomplish this, you would want to force them to require tens of thousands of warheads — and still not be confident of knocking out all of America's nuclear force.[264] The daunting nature of this requirement should assure future American

---

264. The Pentagon believes China is developing intercontinental nonnuclear missiles and may deploy them by 2030. If so, infrastructure nodes (e.g,, electrical generating plants and transformers, key air and sea ports, bridges, communication systems, large computer servers for major companies and financial exchanges, natural gas and oil transmission choke points, etc.) may be targeted. This would recommend proliferating or distributing as much of these assets as possible or hardening or actively defending them.

presidents that they would never have to threaten to "use or lose" America's nuclear weapons to deter Moscow or Beijing.

Unfortunately, strategic deterrence is not simply nuclear. As Israel discovered with Iran's massive April 2024 missile strike, nuclear weapons may not deter enemy states from launching nonnuclear attacks that could have strategic consequences. Also, no nuclear force, no matter how well designed, is guaranteed to neutralize or deter all nuclear threats. As Putin has demonstrated in his war against Ukraine, nuclear-armed states can use nuclear threats to deter nuclear-armed alliances (e.g., NATO) from tipping the military scales against them in nonnuclear wars. With any bad luck, China, Russia, or North Korea may repeat Putin's nuclear dares over Ukraine to stare down the West's will to arm its allies and defend itself against conventional attacks.

In these cases, the proper remedy is more nonnuclear than not. To deter conventional Russian, Chinese, or North Korean attacks, the United States and its allies will need sufficient conventional forces to destroy invading armor, ships, and planes. Increasingly, these forces will consist of unmanned weapons fed with copious amounts of targeting data and fortified with artificial intelligence. This, in turn, will require air and space-based target acquisition and navigation systems to convey vast amounts of information via super secure space, air, and seabed-based broad bandwidth communications networks. The United States will also need to harden and protect these secure networks against attack.

Leverage U.S.-Allied Nuclear Diplomacy

This brings us to the second required strategic adjustment: Employing new forms of nuclear diplomacy to counter Russian and Chinese nuclear threats and complaints against U.S. nuclear sharing geared to drive a wedge between America and its allies.

The conventional wisdom is that arms control and nonproliferation diplomacy are impossible without the cooperation and engagement of Moscow and Beijing. Since Russia and China currently see America and its allies as their enemies, U.S. and allied diplomats cannot expect to reach any major, new arms control or nonproliferation treaty agreements. This has discouraged American and allied diplomats from proposing any bold arms control or nonproliferation initiatives.

Meanwhile, Russia and China have been busy. Russia has proposed "demilitarizing" space, threatening to use its theater nuclear weapons if NATO helps Ukraine directly, and demanding that NATO remove all American nuclear weapons from Europe. Not to be outdone, China has promoted agreements on no first use of nuclear weapons and warned Washington against supplying nuclear submarines to Australia or redeploying nuclear weapons on South Korean soil. China also has refused to engage in "good faith" negotiations to limit or control nuclear arms as required by Article IV of the Nuclear Nonproliferation Treaty (NPT) until China reaches nuclear parity with the United States. Similarly, Russia has frozen arms control talks with the United States until NATO stops supporting Ukraine.

Xi's and Putin's aim is to weaken America and drive a wedge between it an its allies. American and allied diplomats instead are mostly focused on creating new confidence-building measures, arms control understandings, and binding treaties with Beijing and Moscow. As desirable as securing such understandings may be, leveraging the West's comparative military and technical advantages against China and Russia is even more important.

Recently, the Biden Administration proved this point. It asked Japan to demand that Russia back a United Nations Security Council resolution affirming Russia's Outer Space Treaty obligation not to put nuclear weapons into orbit. The resolution had the support of 62 nations, most of whom rely upon U.S. and Western satellites to meet their communications and imagery needs. Russia and China tried to turn this resolution to their advantage by demanding all weapons in space be banned, including ones permitted by the treaty. It didn't work: Russia, caught testing a satellite designed to carry a nuclear weapon, was isolated. It alone voted against the Japanese resolution. China, not wanting to be on the losing side, abstained.

The United States and its allies ought to double down on such diplomacy. An ideal way to do so is to focus on nuclear sharing. Late in 2023, the Pentagon spotlighted[265] China's probable use of its "peaceful" fast reactor and plutonium recycling programs to match or exceed America's nuclear force by 2035.[266] At the same time, China has repeatedly warned Washington not to share U.S. nuclear weapons with South Korea or Japan.[267] Presumably, Beijing does not want Washington to undermine China's nuclear build up by redeploying U.S. nuclear arms to its Asian neighbors.

This hypocrisy presents an opportunity.

From a military scientific perspective, Washington has little to gain from redeploying its nuclear arms to Asia. It originally based U.S. nuclear weapons overseas when it lacked intercontinental nuclear missiles, long-range air and sea-launched cruise missiles, and a reliable intercontinental-range bomber force. By the late 1980s, it had all these modern secure nuclear delivery systems and no longer had to risk placing its nuclear arms in vulnerable bases overseas. As a result, Washington dramatically reduced the number of

---

265. Yuki Kobayashi, "China's Fast Breeder Reactor Operating? Possibility of Accelerating Nuclear Arms Race," *SPF China Observer*, November 30, 2023. https://www.spf.org/spf-china-observer/en/eisei/eisei-detail006.html.
266. Idrees Ali and Phil Stewart, "China likely to have 1,500 nuclear warheads by 2035: Pentagon," *Reuters*, November 29, 2022. https://www.reuters.com/world/china-likely-have-1500-nuclear-warheads-by-2035-pentagon-2022-11-29/.
267. Henry Sokolski, "Let's call China's bluff on nukes and plutonium," *The Hill*, August 23, 2023. https://thehill.com/opinion/national-security/4164909-lets-call-chinas-bluff-on-nukes-and-plutonium/.

weapons it based in NATO countries (from 7,000 to roughly 100) and in East Asia (from more than 3,000 to zero).[268]

South Korea's president, Yoon Suk Yeol,[269] and former Japanese president, Shinzo Abe,[270] have spoken of the possible need for the United States to redeploy nuclear weapons to Japan and South Korea. Washington, so far, has demurred and instead offered increased nuclear planning consultations and exercises. For the moment, this has satisfied South Korea and Japan.

It's unclear how long this will suffice. As Chinese and North Korean nuclear arsenals grow, it is likely Japanese and South Korean interest in acquiring nuclear arms will as well. To satisfy this prospective demand, the RAND Corporation has suggested an intermediate fix: Have Seoul pay to construct nuclear storage sites in South Korea and modernize 100 American tactical nuclear weapons that Washington otherwise plans to dismantle.[271] Washington could then keep these refurbished weapons in the United States for quick deployment if South Korea should ever need them.

Validating this proposal's value, China has already objected to it.[272] This opens a door: Washington should point out that it has no

---

268. Robert Norris, William Arkin, and William Burr, "Where they were," *The Bulletin of Atomic Scientists*, November/December 1999. https://www.archives.gov/files/declassification/pidb/meetings/where-they-were.pdf.
269. Chloe Sang-Hun, "In a First, South Korea Declares Nuclear Weapons a Policy Option," *The New York Times*, January 12, 2023. https://www.nytimes.com/2023/01/12/world/asia/south-korea-nuclear-weapons.html.
270. Jesse Johnson, "Japan should consider hosting U.S. nuclear weapons, Abe says," *The Japan Times*, February 27, 2022. https://www.japantimes.co.jp/news/2022/02/27/national/politics-diplomacy/shinzo-abe-japan-nuclear-weapons-taiwan/.
271. "Options for Strengthening ROK Nuclear Assurance," RAND Corporation, October 29, 2023. https://www.rand.org/pubs/research_reports/RRA2612-1.html.
272. Park Chan-kyong, "Should South Korea 'scare Kim' with US nuclear bombs? 'China and Russia would raise hell'" *South China Morning Post*, May

desire to redeploy nuclear weapons either to South Korea or Japan, but that China's and North Korea's continued nuclear build up produces a clear demand signal for Seoul and in Japan to urge American redeployment of nuclear arms. If Beijing is serious about preventing future U.S. nuclear deployments to the region, China need only reduce the demand signal generated by its and North Korea's continued nuclear build up.

A simple way to do this would be for China to freeze further construction and operation of plutonium recycling and fast breeder reactor plants, which can produce hundreds of bombs worth of super-weapon grade plutonium a year. The international community could verify compliance remotely from space. In exchange, the United States might offer to place a moratorium on any forward deployment of its nuclear weapons to East Asia. This moratorium would last as long as China kept the operation and construction of its "peaceful" plutonium recycling and fast reactor activities frozen. Washington could go further, if it wished, and offer Beijing a regional freeze on South Korean or Japanese fast reactor and plutonium recycling activities and even promise to freeze commercial deployment of American fast reactors.

In Europe, nuclear sharing presents different challenges. In 2023, Putin redeployed Russian nuclear weapons to Belarus. Since then, he has protested American plans to refurbish nuclear storage sites in the United Kingdom and possibly to redeploy nuclear weapons there. Russian officials have also protested Sweden's willingness to station U.S. nuclear weapons in wartime and Polish pleas for Washington to deploy nuclear weapons in Poland.[273] Finally, Russia has demanded the United States remove all its nuclear weapons from Europe and suggested that Russia might consider

---

20, 2024. https://www.scmp.com/week-asia/politics/article/3263120/should-south-korea-scare-kim-us-nuclear-bombs-china-and-russia-would-raise-hell.

273. John Tirpak, "President Says Poland 'Ready' to Join NATO Nuclear Sharing," *Air and Space Forces Magazine*, April 22, 2024. https://www.airandspaceforces.com/president-says-poland-ready-to-join-nato-nuclear-sharing/.

removing its weapons from Belarus only if the United States and NATO "refrained from undermining the security and sovereignty of Russia and Belarus."

The inconsistency of Russia objecting to additional U.S. nuclear weapons deployments to Europe while redeploying Russian nuclear weapons to Belarus is something even China has spotlighted.[274] Washington should do the same. One way to do so is to explain that while the United States would prefer to take its nuclear weapons out of Europe, it can't so long as Russia continues to deploy nuclear weapons there. To cut this knot, Washington should offer to remove what it has deployed in Europe if Russia pulls all theater nuclear arms east of the Urals and allows effective, mutual inspections.

Meanwhile, the United States and NATO should do everything but forward deploy additional U.S. nuclear weapons. This would include building new and refurbishing old nuclear weapons storage sites in Europe, selling new nuclear-capable planes (F-35s) to any NATO member that wants them, and creating nuclear escrow accounts for Poland and any other NATO member willing to pay to have the United States refurbish and store on American soil the tactical nuclear weapons the Pentagon otherwise is planning to dismantle.

Exploit the NPT

Where might the United States best launch these ideas? The very place Russia and China have put the most nuclear diplomatic pressure on the United States and its closest allies — the NPT preparatory review conferences leading up to the next NPT Review Conference in 2026.

---

274. "Remarks by Ambassador Geng Shuang at the UN Security Council Briefing on Threats to International Peace and Security," Permanent Mission of the People's Republic of China to the UN, March 31, 2023. http://un.china-mission.gov.cn/eng/hyyfy/202304/t20230401_11052883.htm.

Reviewing the NPT affords two diplomatic handles. The first is Article IV. It requires NPT members to negotiate in good faith measures to end the "nuclear arms race" — something China has refused to do. Instead, China has insisted that it will only negotiate once it has achieved nuclear parity with the United States. This flies in the face of the clear letter and intent of the treaty.[275] The United States and like-minded nations should drive this home while offering Beijing ways to make good on their NPT obligations.

The second handle is Articles I and II of the NPT, which prohibit the transfer of "control" over nuclear weapons to any state that lacks them. The question here is what does "control" mean? China and many developing nations have long questioned the validity and authority of the agreement reached by the United States and the Soviet Union in the late 1960s. As they point out, Moscow and Washington failed to share their mutual understanding with most of the original signatories to the NPT until after many nations signed, and that, in any case, nuclear sharing violates the treaty's spirit.

Allowing their objections without prejudicing our allies' security, Washington first should hold Putin's feet to the fire by stigmatizing his Belarusian nuclear weapons deployment. To gain the moral high ground on nuclear sharing, Washington should then offer to withdraw US weapons from NATO states if Russia verifiably withdraws its theater weapons locations East of the Urals. Washington could further burnish its nonproliferation credentials by promising not to redeploy its weapons to East Asia as long as China suspends its fast-reactor and plutonium recycling-related activities.

---

275. Henry Sokolski, Thomas Grant, "China: A nuclear nonproliferation deadbeat," *The Hill*, August 30, 2022. https://thehill.com/opinion/national-security/3621641-china-a-nuclear-nonproliferation-deadbeat/.

Be Patient

Would China and Russia immediately agree to these nuclear control proposals? Probably not. Eventually, however, they might. At the height of the Cold War, the United States and the Soviet Union struck agreements to demilitarize Antarctica, ban nuclear weapons testing above ground, prohibit weapons of mass destruction in space, and limit the spread of nuclear weapons.

How was this possible? It helped that Stalin had died and that Khrushchev came into power. It helped even more that several major nuclear crises spooked both sides back to the negotiating table and that, by the mid-1960s, both knew their strategic forces were immune to being knocked out in a first strike.

These past prerequisites may come into play again. Putin's or Xi's grip on power will eventually give way. America now is committed to making itself less vulnerable to strategic attacks. As for the prospects of a major war, they are increasing. With any luck, these developments will produce the very tensions needed to catalyze sound diplomacy. We should get ready.

Seoul's Current Nuclear Weapons Dilemma: What Does History Teach?

Richard Lawless

Overview

President Yoon Suk Yeol's public comments on January 12, 2023, that South Korea may need to acquire nuclear weapons highlighted what an increasing number of South Koreans view as their deteriorating security situation. North Korean direct threats against the Republic of Korea, as well as worries that America's long-standing commitment to "extended deterrence" may be eroding, have only fortified their concerns.

This same theme was echoed and reinforced on February 20, 2023, by the head of South Korea's ruling People Power Party Chung Jin Suk, who expressed serious concerns about the ability of South Korea's current "kill chain" preemptive strike strategy to deter North Korea. In making his case for an indigenous nuclear weapons capability, Chung went beyond President Yoon's earlier comments with his statement, delivered that day to a gathering of his party leadership. "We need to seriously consider developing our own nuclear capabilities if such a response is insufficient." (See Christian Davies, "South Korea Ruling Party Leader Makes Case for Nuclear Weapons," *Financial Times*, February 21, 2023.)

None of these current concerns are new. In fact, all have been components of an internal dialog in South Korea for many years as concern over North Korea's possession of nuclear weapons and

Pyongyang's ability to deliver them with increasing precision and confidence. Beyond this, there are historical parallels with South Korea's first attempt to acquire nuclear weapons in the early 1970s that deserve closer review.

This essay explores Seoul's current security concerns and consideration of acquiring nuclear weapons as a "fix" by reviewing why and how its first attempt to acquire nuclear weapons was made and unfolded. It then uses this history to consider how such a decision on the part of South Korea in 2023 might best be detected, monitored, channeled, mitigated, and managed to inform U.S. policy making.

Given the public nature of this subject, and the degree to which it is being discussed in official and unofficial U.S. and ROK forums, such a decision would likely not result in Seoul again undertaking a covert nuclear weapons development program. (In the past, such an approach was judged to be necessary and unavoidable.) Once decided, such a program would involve fundamental alliance compromises and would alter the structure and mutual commitments of the bilateral security and political relationship.

However unlikely it may appear at this stage, this scenario is highly possible. Once Korean leaders make such a decision, in a decision process that could well occur independent of a dialog with the United States, South Korea and the United States would find themselves challenged to craft a smooth way forward. Korean determination to proceed would meet strong American resistance but would be relevant only up to a point. The end product, a South Korea in possession of its own nuclear deterrent and thus capable of exercising its own security options, would not necessarily doom the bilateral security relationship.

Shades of South Korea's Nuclear Weapons Past

The present situation shares some of the dynamic that existed in an earlier chapter in the U.S.-ROK relationship, a period in which a number of factors injected uncertainties into the security alliance and caused the ROK leadership to doubt the staying power of the United States. The period was the early 1970s and the overall context was America's defeat in and its unseemly disengagement from the Republic of Vietnam. South Korean President Park Chung-hee and his national security team, having committed forces to fight alongside American and South Vietnamese troops in combat there, watched each iteration of the American disengagement with growing concern. It was not so much the matter of the departure itself - as it was obvious that neither the Johnson nor the Nixon administrations had the will to win the war by taking it to North Vietnam- but rather the wholesale abandonment of that country by all elements of the United States body politic.

Congress had weighed in to prohibit U.S. support to the struggling Vietnamese and, emboldened by Kissinger's secret talks in Paris, Hanoi had launched a conventional war that doomed the RVN to destruction. In this same period, we had the Kissinger/Nixon outreach to China and all that portended for established relationships in East Asia, South Korea included. Layered in was the growing impact of Nixon's "Vietnamization" program, begun in 1969, under which the U.S. strove to transfer as much of the fight as possible to the South Vietnamese forces as the American withdrawal from its commitment progressed. The South Koreans saw this for what it was - a component of a policy to reorder the U.S. position in the world. If the Vietnamese were expendable there was no reason in a world of "realpolitik" that Seoul could not be jettisoned.

Park and his close circle had other issues of concern with the United States, and it was a delicate time for the security relationship. The United States kept Park on a short leash in terms of responses to North Korean provocations, to the degree that military actions sustained by Pyongyang over a period of several years failed to elicit any reciprocal response. The attack on the USS Pueblo and the capture of its crew, the concurrent commando attack on Park's Blue House targeting Park for assassination, the carefully-staged shoot down of the unarmed U.S. Navy EC-121 electronics warfare aircraft in international airspace that killed 31 men, numerous ground attacks along the DMZ in which hundreds of South Korean and many Americans were killed, added to Park's frustration.

The consistent U.S. response to these deadly provocations was delayed decision making, overlaid by lame actions once that signaled resignation. The U.S. overall posture was reasonably judged to be ineffectual, sending a consistent message that neither the U.S. nor South Korea would respond in kind. This suggested that the U.S. was a fading ally that had been discombobulated by its failure in Vietnam, at least in the eyes of those in Seoul counting on the United States to stand tall if North Korea invaded in force.

At this same time, the ROK side concluded, in an assessment that was initially resisted by the U.S. intelligence community, that North Korea was doubling down on its forward deployed conventional forces. Although the U.S. came to accept this grim assessment, to Park this reluctance to recognize increased North Korean capacity for offensive operations suggested a weak alliance partner that was unwilling or unable to confront the North Koreans. Park's assumption was that there was a real possibility that the U.S. would tire of its all-in commitment to South Korea, reduce its conventional forces there, and possibly withdraw the only real capability that could deter North Korea from launching a conventional attack - the threat of American nuclear weapons impacting North Korea in the event of war. Park and his team were under no illusions that, when and if the U.S. ground forces departed the ROK, irrespective of any U.S. commitments to the contrary, the U.S. nuclear deterrent would evaporate.

Well along in the process of losing confidence in the sustainability of the U.S. commitment and planning for the day when America could leave South Korea in an exposed position, Park decided in the Spring of 1974 to launch a covert strategic weapons development program. This initiative had four components - a nuclear weapon that could be configured for delivery by a surface-to-surface missile, a ballistic missile system to deliver the same, a chemical warfare capability, and a biological warfare capability. The focus was almost entirely on the first two endeavors, and the issue was the time required to achieve such a capability, or at least the tangible capacity to build and deploy such a system. Park structured his strategic weapons program as a covert undertaking from the overall planning and execution perspective, but proceeding at maximum efficiency required both a military and non-military component. The latter was critical given the inherently visible profile of the nuclear fuel cycle and the nuclear materials that would have to be irradiated and chemically processed, that is, spent fuel reprocessed, to yield the fissionable plutonium required by the warhead itself.

The fuel cycle activity was assigned to the government-controlled Korea Atomic Energy Research Institute (KAERI), a modest organization possessing a couple of small U.S.-supplied research reactors. Empowered by the parent government ministry to move out smartly and given a budget to do all that its researchers could only have dreamed of undertaking in past years, KAERI managers got very busy very quickly to contract for a nuclear research reactor, a nuclear fuel reprocessing plant, fuel fabrication facilities, and associated labs. Under then current international nuclear cooperation standards, all of this was totally acceptable as a complimentary activity to South Korea's ongoing nuclear power generation program.

By mid-1974, KAERI was building itself out as a top-flight research organization, aggressively recruiting from abroad to bring home experienced Korean nuclear engineers. In the space of months, KAERI found itself more than capable of inducing all of the expertise required to create a new research center capable of delivering the fissionable material the Park team needed for its weapons program. With these elements of the overt program on the way, the only issue was the time needed to put all of the piece parts in place. A reasonable goal would be KAERI operating a complete fuel cycle capable of generating quantities of plutonium within four years, that is, by 1978. This was an ambitious but not unrealistic goal given the quality of the researchers and facilities which Park's spare-no-expense funding was delivering.

On the covert side, all responsibility devolved into the capable hands of the Ministry of Defense's Agency for Defense Development (ADD), a relatively new organization which had been created to reverse engineer and develop conventional weapons systems for the ROK military. ADD's mission was to actively support the advancement of Korea's civilian industrial sector committed to the domestic production of weapons and munitions for self-sustainment and export purposes. Under Park's plan, ADD would create a nuclear weapons design group, as well as a missile development team that would engineer the existing U.S. Nike Hercules air defense system to provide the initial strategic weapons delivery vehicle. In parallel, the development of more capable missile systems was initiated to perfect a later phase delivery capability that could range all of North Korea. Extensive funding for an entirely new ADD research center was buried deep within another Park initiative - a special multi-year military upgrade allocation named for a scholar-warrior of a previous dynasty. The "Yulgok Plan" was semi-visible to the U.S. side but allowed plenty of space for the covert funding for all of ADD's activity that was deemed protected from U.S. military and civilian visibility, not to mention interference. Park's planning was intuitive and comprehensive.

Very fortunately, although the South Korean side made extensive attempts to conceal all of the constituent elements of the program from the United States, that is, both the ADD covert weapons design activity and its link to the complimentary KAERI fuel cycle activity, the CIA was able to penetrate the covert program. This penetration occurred within months of the Park decision, and while it took several months to expand the penetration and understand all its working elements, the CIA presence in Seoul was able to provide the U.S. Government with a complete view into almost all of the ongoing work. In this clandestine reporting activity, the Agency found itself literally tracking progress as ADD and KAERI managers strived to assemble their teams and build out the program's covert and overt facilities.

Along the way, particularly in the early months of the Agency's clandestine source reporting, there were issues at home within the U.S. intelligence and policy communities. At CIA Headquarters, analysts doubted the veracity and quality of the early intelligence and had to be convinced that their agency colleagues in Seoul had it all right. Back at the Langley flagpole, a shortcoming that too frequently entered the process came to the fore when superiors overrode operational judgements made in the field on a given reporting area. This top-down approach wherein home base could and often did shut down field station reporting by assigning a grade to collected intelligence which conveyed that "Washington is not interested." Too often, and here the author speaks from first-hand experience of five years in Seoul, we were told that information which we in the field deemed to be important to policy makers, was not valued at home base. Such evaluations came either from analysts who knew it all and were not to be challenged, or from disinterested policy consumers.

At policy levels, at that point in time, the U.S. government did not yet have a comprehensive or even coherent non-proliferation policy from which to work. The entire U.S. Government was caught in its own dilemma coming out of the Eisenhower era "Atoms for Peace" initiative that advocated the world-wide adoption of nuclear power and all the accoutrements that attended such, to include the fuel cycle itself.

Other developed nations which had embraced their own nuclear future, to include Canada, the United Kingdom, France, Germany, and others, were each determined to promote with third-party sales and technical assistance their respective technologies and solutions. In terms of nuclear technology, in the early 1970s we were experiencing a "wild west land rush" among a host of would-be nuclear suppliers competing to sell various reactor designs and differing fuel cycle methodologies. This was occurring with almost no regard for the intentions of nations to which these transfers were directed. Soon-to-be nuclear proliferators were the direct beneficiaries of this surge, including India, Pakistan, Israel, South Africa, and Taiwan among others. South Korea simply got in line and found a wealth of vendors, all eager to sell sensitive technology, often with the quid pro quo that sensitive technology buyers would commit to purchase nuclear power reactors nominated by the fuel cycle vendors.

The agency's team in the field in South Korea was vindicated when it was shown that its clandestine reporting had been extremely accurate from day one. Human source reporting was collaborated as buildings came out of the ground where reporting stated they would be built in remote locations to support the covert program. As foretold, contracts were signed with foreign vendors behind closed doors for KAERI fuel cycle facilities, the existence of which was denied to U.S. officials. In Langley, the analysts soon fell into place, made the correct assessments as to Park's intent and the capability of KAERI and ADD to deliver, and from this informed the policy makers. The issue was quickly elevated to Henry Kissinger by his team and he immediately discerned all the implications of the Korean initiative. At this point, seized with the magnitude of this challenge, the State Department and the National Security Council moved expeditiously to craft and execute an aggressive approach designed to deny Park his strategic weapons program. But gaining the upper hand with Park and resolving the issue in a manner that protected the alliance with no collateral disruption to the regional security order would be a year-long hard slog.

Understanding Park's calculations, namely the factors that compelled him to take such risks with the overall U.S.-ROK relationship and with the alliance, was critical. Human-source intelligence allowed the U.S. to confirm that Park was not only seized with the need for the program but very confident that the plan to deliver this capability was feasible. It seemed that Park intended to undertake both the overt component (securing an indigenous nuclear fuel cycle) and the covert companion activity knowing that it could well be discovered by the U.S. and that, when a confrontation did occur, he could still deal with that event to his and to South Korea's advantage.

The author here speculates with hindsight that, in a medium-case scenario, Park may have calculated that if the U.S. did not detect his strategic weapons program until it was down the road a couple of years, with the nuclear fuel cycle in place and weapons designs and missile systems being perfected, he might have advanced his gambit to a point where he could compel the U.S. to accept his indigenous program and accommodate it within the alliance. In a worst-case scenario, again having arrived at the two-year point, if the U.S. then discovered the program and demanded that Park shutter it, he would be in a position to bargain and trade that concession for improved strategic assurances.

Not beyond the range of possibility would be a ROK demand that the U.S. allocate tactical nuclear weapons to the control, either exclusively or under a dual-key arrangement (a la the NATO nuclear sharing program), of the ROK military. In sum, Park's move to initiate the Yulgok 971 project was a very ambitious, high risk but calculated gamble, and as such it was consistent with the inherent "Korean risk taking" mentality. In other words, whatever the outcome, Park's game to quickly achieve a sovereign strategic weapons capability that would provide his nation with an ultimate deterrent against North Korean, or Chinese aggression, was worth the candle.

In the case of the earlier Korean nuclear situation, by sustaining a maximum effort to collect well-sourced and timely-presented intelligence the U.S. policy makers were able to track the results of the U.S. approaches designed to derail the activity. This visibility into the discussions among Park's inner circle allowed the U.S. government to tailor its approach to include leveraging the differing priorities among Park's advisors. Through this, one was able to expose to Park inconsistencies in the promises made by his nuclear team, and raise issues with the Blue House inner circle which we could confirm had not been thought through. One such question that was selectively put to this group was "Have you all developed a plan for the control of strategic weapons once the ROK possesses them?" This was certainly a valid question in the context of Park's one-man rule and it resonated.

We were also fortunate that there existed several very senior and influential Korean officials who did not share Park's desire to develop an independent nuclear deterrent, at least at that time. These men, many of whom were the leading lights of their nation's dramatic economic development as skilled technocrats, immediately recognized that Park's clandestine program was a very bad idea from almost every angle imaginable. Once these individuals were carefully brought into the equation, several individually or collectively resisted and helped bring the program down. Once again, the U.S. ability to selectively share detailed intelligence with these individuals carried the day. The fact that we knew it all, or almost all, and were able to convey this, was decisive.

Still, a Close-run Thing

There was absolutely no doubt among any of the U.S. officials involved in this issue during the 1974-1978 period that there was real concern on the part of the ROK leadership that the U.S. could well abandon the ROK. Close on the devastation delivered by the Korean War, the fear of abandonment, for whatever combination of reasons, be they logical or whimsical, was real and ever present. We were fortunate to acquire the protected information we did secure when we did and to possess the will to apply unrelenting pressure on all of the involved parties to force the issue. The latter included not just Park's regime but also Canada, the United Kingdom, and France once we appreciated the role that each was playing in our contest of wills.

As if to drive home this concern and keep it at the fore in the ROK, the United States elected Jimmy Carter as its president in 1976. Within weeks of Carter taking office in early 1977, his administration would stun U.S. policy makers and shake the U.S.-

ROK alliance to its foundation. Carter made it known that he was determined to withdraw U.S. forces from the ROK and scale down the security guarantees there, the implication being that the ostensible strategic guarantee that had existed as a companion to the presence of conventional U.S. forces there was in doubt. President Park and his inner circle soon came to regret the decision to abandon their strategic weapons program as the reality of the Carter presidency asserted itself a few short years after the U.S. forced the shutdown of that covert weapons program.

Had the U.S. failed to detect and dismantle the ROK strategic weapons program when it did, and had that program still existed in 1977 when Carter arrived on the scene, the eventual outcome could well have been quite different. It is probable that different calculations would have driven different decisions in the Blue House if the facilities and mechanics were then still in place to allow the weapons system to be built. Almost certainly, given the drama that attended the Carter years and the impact of his obsession with the U.S. presence in South Korea, a ROK strategic weapons program would have been sustained if not accelerated. In the final accounting, both nations were fortunate that the entire issue was resolved when it was and that minimal damage was delivered to the relationship and the alliance.

An Intelligence Approach in 2023 to Non-Proliferation in Northeast Asia

In 2023, the dynamic among Japan (as a U.S. ally with embedded strategic security guarantees), the ROK (ditto albeit with nuanced differences), North Korea, and China (the latter in both the Taiwan and first island chain context) is historically fraught with tension and unpredictability. In policy discussions in years past, the author has compared this region to a nuclear reactor in which the United

States plays the role of the control rods vertically inserted to moderate the chain reaction among the fuel elements, allowing the system to generate heat while preempting a criticality run-away that would destroy the facility and contaminate all around it. In the current case, we have disparate fuel elements inserted and active, some of which are beyond the ability of our control rods to moderate, with the certain prospect that ever more complex fuel elements are being crafted, soon to be introduced into the mix.

At some point, certainly within the next few years, the U.S. role as the ever-challenged manager and control rod provider for this East Asia nuclear reactor will no longer be sustainable. This could occur as a result of simple fatigue with the mission, or part of a broader national retrenchment that abandons certain too-difficult obligations far from home, or because the U.S. has come to reluctantly judge that this effort is beyond its reasonable capacity to sustain even if it wished to do so. In such a case, the U.S. may be forced to distance itself from the Northeast Asian reactor as it begins to go critical, and leave the system to its own ends. As one senior U.S. diplomat once said of the U.S. commitment to do its level best to deliver on its stewardship of Korean and Northeast Asia stability as beyond America's writ, referencing a Chinese proverb, "It may be time to just poison the shrimp and burn the fields and depart."

Howsoever the U.S. geostrategic position in Northeast Asia evolves, and in that evolution what solutions we find or fail to find regarding penultimate security guarantees to the ROK, as well as to Japan, the experiences of America's non-proliferation exercise in the 1970s suggest some must-have elements of an action plan that keeps U.S. policy makers on top of developments in the region. A key theme for all intelligence collectors, be they clandestine collectors working the streets or imagery analysts bent over a photo table in a non-descript building just outside the Beltway, is the mantra of "no surprises."

Detailed timely knowledge allows preemption and, in fact, such information is the sine qua non for all policy formulation and execution. The failure to collect aggressively and work objectively with that collected product, discarding presumption and prejudice, condemns national leadership to a come-from-behind scramble that too often gets the solution wrong. Put another way, grand intelligence failures, of which the United States has had far too many, are acceptable if the missed event is a coup in West Africa. But in the category of nuclear issues, with the Cuban Missile Crisis as our poster child for failure writ large, there is no margin for poor performance. This is particularly so when we have multi-billion-dollar intelligence budgets feeding the trough from which a score or more of collection entities feed.

Intelligence Considerations for Future Practitioners

Revisiting the former non-proliferation adventure with South Korea and venturing a guess that Seoul will sooner rather than later conclude that it has little choice but to again seek the ultimate security guarantee, the following considerations related to intelligence collection and processing are suggested:

- Do not assume that the recently resurrected U.S.-ROK dialog on strategic assurance or nuclear reassurance will carry the day, at least beyond the next few years. It is probable that North Korea, in adjusting its confrontational strategy, will continue to emphasize provocations designed to embarrass and degrade the ROK leadership and undermine the credibility of the U.S.-ROK security alliance. In such a case, every nuance of the South Korean reaction will require close attention.

- If the U.S. elects to deepen its strategic guarantee to Japan, possibly with the stationing of INF-class conventional and nuclear missiles, either on land or at sea to protect the Japanese home islands, the ROK will expect and accept nothing less. Failure to secure the same level of U.S. protection and commitment will trigger a go-for-broke program to build and field a nuclear weapons deterrent. South Korea's ability to realize such a program should not be underestimated, nor should the U.S. ability to constrain such a program be overestimated. Such capabilities for ROK breakout require the attention of intelligence collectors, of both the covert and overt varieties, as well as a constant updating of assessments by analysts dedicated to tracking this target.

- While North Korea's ostensible end goal is the reunification of the peninsula under Pyongyang's control, the near-term, must-have objective of the Kim regime is a direct relationship with the only nation on the face of the planet that can assure its survival, namely the United States. The Kim dynasty would be willing to sacrifice millions of South Koreans to achieve this end. If the next nuclear weapon test successfully demonstrates that a neutron weapon is in hand, South Korea has reason to tremble. Meanwhile, already holding Japan proper and U.S. bases there hostage to its missile forces, Pyongyang will accelerate its ICBM program to credibly threaten the U.S. homeland. A North Korean scheme that imagines that it will be able to dissuade and deter any U.S. strategic response to a substantial conventional or nuclear strike on

Japan or South Korea or both is increasingly in the realm of the possible. The ROK and Japan have concluded as much.

- Expect no assistance from China, Russia, or any other actors as long as North Korea's aggression is directed at America and its Asian security partners. We need to actively collect to determine the attitude of the Chinese leadership as well as its military planners as North Korea becomes more disruptive to the regional order. China probably expects that, as Pyongyang's nuclear capability grows, it will be even less restrained than it is already, and increasingly unresponsive to Chinese attempts to influence Pyongyang's behavior. While we should assume that China will remain standing on the sidelines as North Korea runs its playbook, we should not assume that there will not be a breaking point for China. The possibility of such is too important to miss and we need to collect intelligently with such an eventuality in mind. Similarly, on the South Korean nuclear front, it is likely that a nuclear South Korea would be an inherently unacceptable development for China. Thus, we need to track China's assessment of and reaction to the dialog that has begun in Seoul over a possible nuclear future.

One Korean peninsula nuclear futures scenario which has not received adequate attention but requires additional focus is the potential for a North Korean regime collapse and the consequent fate of Pyongyang's nuclear arsenal. Prudent planning suggests that the ROK would desire to seize and incorporate some or most of that capability into its own force structure in the event of such a collapse. Even the best-intentioned leadership in Seoul would probably find itself under substantial domestic pressure to gain and sustain some degree of control over such "national" assets. Covert collectors and analysts cannot ignore the potential for such a Black

Swan event, even if the analytical elite judge it to be unlikely. Because the national security implications of such a fast-moving event are serious given the regional reordering that such a situation would occasion, policy makers need to be informed by balanced assessments and game out these scenarios. Critical components of such an assessment will only be derived from the ROK mission planners so involved, not to mention their Chinese counterparts and even Japanese counterparts.

The potential for a nuclear South Korea in this decade remains high and worthy of deliberate attention and planning. Solid intelligence will be needed to inform the analytical process from which realistic policy scenarios will be crafted. All-source intelligence, to include human intelligence reporting that provides intent and context, will perform an essential role. An approach that fails to put such a comprehensive collection effort in place today will leave policy makers with a come-from-behind scramble to catch up to events, leaving little margin for error given the velocity at which decisions will be made.

One additional thought, overriding some of the above, but reinforcing most of it, is as follows: There is the case that the U.S. recognizes that the denuclearization of North Korea is an impossible goal however diligently that is pursued and concludes that it is time to either walk away from the "vortex" or opt to explore a solution that may seem inherently foreign to its best interests as well as to the entire non-proliferation regime. In that realization, U.S. may manage to bring both South Korea and Japan along with it to recognize North Korea as a bona fide nuclear weapons state and deal with it formally and frontally and objectively as such. There is little doubt that Pyongyang would welcome such an engagement, in company with the recognition of its statute that such an understanding would confer. It would also do everything possible to insist on an elevated North Korea-United

States partnership that relegates South Korea, Japan, and even China to a secondary status in Northeast Asia.

If the U.S. sought to explore such a solution, the goal being to create and sustain a nuclear modus vivendi of sorts in the region, there will be a range of compromises in the works that impact, possibly irritate the other involved nations, with secondary effects that would need to be anticipated, modeled, contained, and managed. It is obvious that none of this could occur absent an aggressive and comprehensive intelligence collection and analysis effort. In such a policy move, or even the consideration of the same, that capability would have to be in place, collecting and providing policy makers with quality intelligence from the outset.

About the Authors

The Hon. Christopher Ford is a Visiting Fellow at Stanford University's Hoover Institution and a Distinguished Fellow at the National Security Institute at George Mason University's Antonin Scalia School of Law. From February 2021 until January 2023, he worked at the MITRE Corporation, most recently as a MITRE Fellow and the founding director of the Center for Strategic Competition. Before coming to MITRE, Dr. Ford served as Assistant Secretary of State for International Security and Nonproliferation, also exercising the authorities of the Under Secretary for Arms Control and International Security. In prior Executive Branch positions, he has served as Special Assistant to the President and Senior Director for WMD and Counterproliferation at the National Security Council, U.S. Special Representative for Nuclear Nonproliferation, and as Principal Deputy Assistant Secretary of State for Verification and Compliance. Outside the Executive Branch, Dr. Ford has served as a senior staffer on five Senate committees – Foreign Relations, Banking, Appropriations, Intelligence, and Governmental Affairs – as well as an intelligence officer in the U.S. Navy Reserve, a practicing lawyer, and a Senior Fellow at Hudson Institute. A *summa cum laude* graduate of Harvard, Oxford (as a Rhodes Scholar), and the Yale Law School, he is the author of three books and numerous articles and papers.

Dr. Thomas D. Grant is a Senior Fellow of Wolfson College in the University of Cambridge; a Fellow of the Lauterpacht Centre for International Law at that university; and a Visiting Fellow of the National Security Institute at the Antonin Scalia Law School. He has worked for close to thirty years at the intersection of international law, international security, public policy, technology, and private practice. He numbers among a small handful of Americans who have served as legal counsel in cases before the International Court of Justice, Permanent Court of Arbitration, investment treaty tribunals, commercial arbitration tribunals, and under the law of the sea dispute settlement system. For the past twenty years, he has advised companies and governments at CEO/general counsel and head of state/head of government level on strategy in negotiation settings and in litigation and arbitration. He has been a United States designee to the Permanent Court of Arbitration roster and a United Kingdom alternate designee to the arbitrators roster under the EU-UK Trade and Cooperation Agreement. From 2019 to 2021, Dr. Grant served as Senior Advisor for Strategic Planning in the Bureau of International Security and Nonproliferation in the U.S. Department of State. Leading academic journals and publishing houses have published Dr. Grant's writings on defense and security, nuclear nonproliferation, sovereign immunity, banks and international financial crime, international maritime delimitation, state succession, customs and trade unions, the recognition of states and governments, and the conceptual foundations of machine learning (AI), among other topics. His work has been cited with approval by appellate and final instance courts in the United Kingdom and Sweden and by party counsel in several forums, including before UNCLOS Annex VII tribunals. Dr. Grant also writes from time to time for the general public; articles of his have appeared in the *Wall Street Journal, Washington Post, Los Angeles Times, Law360, Bloomberg, The Hill, The National Interest, RealClear Defense,* and other outlets. At the Department of State, Dr. Grant, reporting to Assistant Secretary Christopher A. Ford, led the implementation of the 2017 National Security Strategy across the projects and programs of the Bureau of International Security and

Nonproliferation, including in regard to counter-China initiatives and coordination with allies and partners on export controls on sensitive technologies. In private practice since leaving the State Department, Dr. Grant has collaborated with former ITC Commissioner F. Scott Kieff on a number strategic consulting engagements to startup companies as well as established market players in patent infringement matters before the U.S. International Trade Commission (ITC). Dr. Grant has taught at the BA, LLM, and PhD levels at the University of Cambridge, Catholic University of Lille, University of St. Gallen, and University of Geneva. Tom received his BA from Harvard, *summa cum laude,* where he concentrated in modern European history and was Junior Phi Beta Kappa (top twelve students at end of junior year); his JD from Yale; and a PhD in international law at the University of Cambridge, where he studied on a Fulbright grant. He was born in Gettysburg, Pennsylvania and raised in North Attleborough, Massachusetts, where he attended public schools K through 12.

Eldon Greenberg is a retired partner at the law firm of Garvey Schubert Barer. Previously he served as counsel to the Nuclear Control Institute in Washington, DC. He also served as general counsel of the National Oceanic and Atmospheric Administration (NOAA) and as deputy general counsel of the Agency for International Development in the administration of President Jimmy Carter.

Richard Lawless is a Founder and Principal of NMV Consulting, LLC. Richard also serves as the Chairman Emeritus of the Board, Texas Central Partners, LLC, a partnership that is deploying high-speed rail technology on the 240-mile Dallas-Houston, Texas corridor. Finally, Richard is a co-founder and Chairman of mobile technologies development company, Modo Labs, Inc. Richard served the United States government for over 20 years, most recently as Deputy Under Secretary of Defense for Asian and Pacific Security Affairs. In this capacity Richard was responsible for formulating U.S. security and defense policy in the Asia-Pacific region, including East, South East and Central Asia. He retired from this position in July 2007. Prior to his appointment at the Department of Defense, Richard co-founded and served as the Chairman/CEO of U.S. Asia Commercial Development Cooperation from 1987-2002. Mr. Lawless served as a career employee of the Central Intelligence Agency from 1972 through 1987, serving in Washington, D.C. and various postings in the Far East and Europe. He specialized in subjects related to high technology, nuclear proliferation and Far East security issues. Richard is a graduate of Bradley University's School of International Studies (B.S. International Relations, Magna Cum Laude) and the Defense Language Institute, Monterey, California (Korean language program). In addition to the January 2023 publication of "Hunting Nukes", a second book, "Nightstalkers- The Wright Project and the 868th Bomb Squadron in World War II", will be available from Casemate Publishing (UK) in March 2023.

Henry Sokolski is executive director of the Nonproliferation Policy Education Center and teaches graduate classes on nuclear policy at the University of Utah and the Institute of World Politics. He was recently appointed Senior Fellow for Nuclear Security Studies at the University of California at San Diego. Previously, he worked in the Pentagon as Deputy for Nonproliferation Policy, as a consultant to the National Intelligence Council, as a member of the Central Intelligence Agency's Senior Advisory Group, as a Senate military legislative aide to a member of the Senate Armed Services

Committee, and as a special assistant on nuclear energy to the chairman of Senate's Tennessee Valley Authority oversight committee. He served on two congressional commissions on the prevention of WMD proliferation and has authored and edited numerous volumes on strategic weapons proliferation issues, including Best of Intentions: America's Campaign against Strategic Weapons Proliferation and Underestimated: Our Not So Peaceful Nuclear Future. Mr. Sokolski earned his graduate degree in political science from the University of Chicago.

Andrea Stricker is a research fellow at FDD where she conducts research on nonproliferation, Iran, North Korea, and other security policy topics. She is an expert on nu clear weapons proliferation and illicit procurement networks. Andrea has extensively researched Iran's nuclear program, including its history, proliferation efforts, and diplomatic agreements, particularly issues relating to the Iran nuclear deal and Iran's Nuclear Archive. She has also written in-depth studies of strategic commodity trafficking and proliferation financing and advised federal prosecutions on illicit procurement cases. Her other areas of interest include non-proliferation regimes and related United Nations resolutions, illicit finance, Middle East security dynamics, sanctions, and U.S. foreign policy. Prior to joining FDD, Andrea spent more than twelve years at the Institute for Science and International Security authoring publications that integrated technical and policy research to make assessments and recommendations on preventing nuclear proliferation. Preceding that, she worked with the International Rescue Committee to enhance her focus on refugee issues and human rights atrocities as they pertain to countries and regions in conflict and a lack of rights to self-determination. Andrea is the co-author of such books as Taiwan's Former Nuclear Weapons Program, Revisiting South Africa's Nuclear Weapons Program, and The Peddling Peril Indices (PPI), and has contributed chapters to The Plutonium Handbook, Iran Primer, and Preventing Black Market Trade in Nuclear Technology. Her publications and analysis have appeared in The Washington Post, Wall Street Journal, New York Times, Los Angeles Times, Chicago Tribune, Guardian, London Telegraph, World Export Control Review, Washington Quarterly, NK News, National Interest, Voice of America, Atlantic, Newsweek, TIME, Science Magazine, and other domestic and foreign media outlets.

Appendix: Nuclear NPT War Games

Prescription for Military Paralysis: Wartime Reactor Meltdowns

For more than a year, nuclear experts have wrung their hands about the risk of radiological releases from the Zaporizhzhia nuclear power plant in southeastern Ukraine and how best to prevent them. More than 15 months into the war, though, Russian attacks against Ukraine's nuclear plants have released no radiation. This may be no accident. So far, Russian President Vladimir Putin has avoided destroying Ukraine's nuclear power plants. Otherwise, they would have been demolished long ago. Instead, Putin has aimed to damage them and Ukraine's electrical supply system as part of a larger effort to erode Ukrainian morale.

His strategy is unlikely to be a one-off. North Korea and China also have "wayward provinces"—South Korea and Taiwan, respectively. And they have long-range missiles too. Beijing and Pyongyang have considered targeting reactors. How these countries might, if at all, follow Russia's example depends on what they make of Putin's current assaults against reactor sites.

Reactors in Warzones:

When he launched his full-scale invasion on February 24, 2022, Putin hoped Ukraine would immediately surrender. His initial aim wasn't to disable Ukraine's reactors or electrical supply systems but to seize them. And he did: On the first day of the invasion, Russian forces took control of the Chernobyl nuclear plant, and in early March they seized the Zaporizhzhia nuclear plant. Russia's invasion, however, quickly stalled. As a result, Putin changed his strategy: He directed his military to shell the plant sites and electrical power nodes critical to powering the plants' coolant pumps and safety equipment. Russian agents also kidnapped and terrorized plant workers, which jeopardized Zaporizhzhia's safe operation. In addition, Putin stepped-up attacks on the rest of the Ukraine's electrical supply system to frighten the Ukrainian population further, undermine its will to resist, and possibly destabilize the entire grid, including the nuclear portions of the electrical supply system. So far, these efforts have had mixed results: Some senior officials in NATO countries have been rattled (fearing radiation leaks and military escalation); the Ukrainians, however, have not.

Certainly, Russia's willingness to take advantage of the military vulnerabilities of nuclear sites in Ukraine has set a precedent. It is unclear if any other nation would make the mistake Russia did in assuming that it could easily seize and hold an adversary's nuclear facilities at the very outset of hostilities. If not, they might move to Russia's second stratagem of militarily holding the electrical supply system and its nuclear plants at risk right away. How might such a war proceed? Might Russia aim to knock out the grid, strike Ukrainian nuclear plants, and risk major radiological releases? Might it target NATO reactors (which could include more than 50 U.S.-promised plants in Poland, Romania, and Ukraine by 2037)? How might Ukraine, the United States, and NATO members respond?

To answer these questions, the Washington-based Nonproliferation Policy Education Center (NPEC) designed and hosted a wargame. The game assumed Russia will re-invade Ukraine 15 years from now—in 2037—when both sides will have substantial numbers of long-range, accurate missiles and drones. It also assumed Ukraine and Eastern NATO countries will have new reactors of U.S. design on their territory. In November and December of 2022, NATO officials, American hawks, American doves, Ukrainians, Romanians, nuclear experts, U.S. military officials, and Polish experts were all tapped to prepare, critique, and play remotely over a two-week period.

The game's play revealed how the uncertainties and dangers of military attacks against nuclear power plants can paralyze decision-making and fundamentally alter the course of wars. The military disruptions these uncertainties introduce may far outstrip the safety issues any reactor radiological release might otherwise present. The game's play revealed three reasons why.

The U.S. and Its Allies Are Unprepared:

Overseas adversaries can easily target allied or friendly nuclear power plants in ways that the United States and its allies are unprepared for. What was stunning throughout the game's play was the reluctance of the players—other than those representing Ukraine and Poland—to act even after Russian military assaults were made against nuclear power plant sites in Ukraine and NATO countries. The United States team, for example, waited and then failed to extract U.S. personnel at reactor sites that Russia had hit in Ukraine and that were leaking radiation. Only after Russian missiles had induced a loss of coolant at one of Ukraine's Khmelnitsky Westinghouse reactors and threatened to do the same to nuclear power reactors in NATO countries did the NATO team take decisive action. This consisted of supplying the targeted plants in Ukraine, Poland, and Romania with active defenses and auxiliary emergency cooling equipment.

In the game, nuclear expert team members gave contradictory advice to each of their teams about how well any of these reactors would fare against aggressive military assaults. This was unexpected but turned out to be significant. Radiological leaks were detected but assessments of these leaks' implications for public safety were only hastily made after the reactors were hit. These assessments also varied widely and were responded to quite differently by each team. The Ukrainians ordered a massive evacuation after Russia struck one of its reactors; Poland, whose reactor site also was hit but was releasing no radiation, took no public safety steps until they detected a radiation cloud from Ukraine drifting over Polish territory.

Although war planners prefer to devise precise, proportionate diplomatic, political, and military responses, this is difficult to do amid ongoing attacks against nuclear plants. The reason why was made evident in the game: The nuclear experts in the game rendered very different assessments of what was happening and how dangerous the assaults on plants might be to the surrounding population. There was a tendency among NATO members, who wished to stay out of the fight, to downplay the safety implications. It was just the opposite among the countries at greatest risk of being painted with radiation. This suggests that such "differences" in threat perception might not be quickly resolved through some technical nuclear forensic assessments of events. The creation of international norms or nuclear safety zones in war zones, meanwhile, may be desirable but are extremely difficult to attain. As such, the risks and benefits of adding new nuclear plants in high-risk war zones must be reassessed.

Reactor Attacks Can Paralyze Allied Responses:

The hesitation of the United States in responding to military assaults against friendly countries' reactors can risk near-fatal fracturing of U.S. security alliances. In the game, NATO countries closest to the fighting (e.g., Poland) wanted to join Ukraine in conducting deep strikes into Russia against key staging bases that were launching attacks against Ukraine's reactors. Initially, some NATO countries were sympathetic to Ukraine striking Russia. All NATO members were concerned that matters might escalate and spill over into NATO territory. As a result, NATO was ready to invoke Article 4 of the NATO treaty, which authorizes NATO members to bring issues of concern to the attention of the organization.

Eager to avoid direct military contact with Russia, however, key members of NATO decided to manage Poland's desire to back Ukrainian strikes against Russia by invoking Article 5. NATO did this less to support military operations (much less to attack Russia) under Article 5, as to deter any independent action Poland might otherwise take against Russia. In the game, the tactic worked: NATO members, including Poland, were deterred from striking Russia. This tactic, however, failed to deter Kyiv. Ukraine unilaterally struck airbases deep in Russia. This action only further amplified the different concerns of NATO members near the action and those of members located farther back.

Such alliance strains can only be addressed in one of two ways: The reactors either must be defended actively or passively so well that radiological releases and electrical failures appear nearly impossible, or alliance war plans and responses must be devised and agreed to in advance and be sufficiently dramatic to deter such attacks. Neither will be easy. As for developing tailored deterrence strategies, the most relevant analogy here may be "pre-planning" to deal with nuclear weapons attacks—a vexing, dubious undertaking at best.

Legal Disagreements about Reactor Attacks:

Attempts to settle the question of whether military assaults against nuclear plants constitute war crimes or if subsequent radiological releases qualify such attacks as nuclear weapons use can themselves become significant wartime distractions. In the game, Ukraine insisted that Russia's attacks against reactors constituted an actionable war crime under Protocol 1 of the Geneva Convention. Ukraine and others also claimed such assaults constituted first use of nuclear weapons.

These claims divided NATO players. They subsequently not only delayed actions critical to waging the war, but also prompted Ukraine to act unilaterally in an escalatory action, firing missiles deep into Russia without NATO's support. This is worth avoiding. Both NATO members (except the United States) and Russia have ratified Protocol 1 of the Geneva Convention, which specifically discourages assaulting nuclear power generating stations. Yet current U.S. legal guidance regarding Protocol 1 is murky. Although the United States is obligated as a signatory to Protocol 1 to avoid attacking nuclear power reactors, Pentagon lawyers insist U.S. commanders should ultimately be free to attack these plants if they think it is necessary. It would be helpful if the U.S. view was clarified and brought into line with the strong presumption of U.S. allies against making such attacks.

Yet another divisive issue is what constitutes first nuclear use. In the game, European NATO members sympathized with Ukraine's contention that Russia's "intentional" attacks against nuclear plants that consequently released radiation should be considered a "use" of nuclear weapons. The United States ignored this assertion. Yet another unresolved legal question is whether or not radiation that contaminates NATO soil from an intentional Russian attack of Ukrainian reactors should constitute an actual attack on NATO and, therefore, demand an Article 5 response. The players were briefed on this point but chose not to play it. Here, again, some European NATO public officials have supported the idea, whereas the United States has taken no position.

Wargame Format:

The wargame consisted of three moves. The first began in 2037. Putin's successor launches a second invasion of Ukraine, and the Russian military assaults and occupies the four-unit Khmelnitsky plant, which now has been expanded to include two U.S.-built reactors in addition to the two Soviet-design VVER reactors. In the second move, the situation escalates and several missiles explode in the parking lots of nuclear power plants in Romania and Poland and hit several supporting emergency diesel generators. The game's last move was a "hot wash" in which the group discussed the simulation and the players' key findings.

Wargame participants were organized into three teams representing the United States, Ukraine, and NATO-EU nations. The control team oversaw communications, managed the scenario, and represented Russia. Teams responded to the crisis, communicated with other teams to gather information, negotiated, and created a response strategy and contingency plans.

Move One:

It is 2037. Putin is dead. His successor, frustrated by the "forced, unjust" armistice reached with Ukraine in 2024, attempts to complete Ukraine's absorption, launching an attack against Ukraine's southern and western salients. Westinghouse has completed two of its promised U.S. reactors in western Ukraine at Khmelnitsky, which are operated with the assistance of U.S. technicians. Additional U.S. and South Korean power reactors have been built as promised in Poland and Romania and are now on line.

After several weeks of fighting, Russian forces assault and occupy the Khmelnitsky plant and garrison missile strike forces at the reactor site. The Ukrainians precisely target Russia's missile units at Khmelnitsky using weapons the United States has shipped to Ukraine. Russia protests publicly, demanding NATO cease supplying such weaponry through Poland and Romania.

Meanwhile, in a repeat of the tactics Moscow used in 2022 against Zaporizhzhia, Russia fires missiles knocking down several power lines into the Khmelnitsky plant. This threatens the continued reliable supply of external electricity to the plant, which is needed to prevent the reactors' cores and spent fuel pools from overheating and releasing radiation.

Spooked, nearby Romania and Poland (both NATO and EU nations) urge their populations to avail themselves of stockpiled iodine pills. Meanwhile, Russia and Ukraine blame each other for targeting the felled power lines. Then, a missile fells the last external power line connected to the plant, forcing it to run on its emergency diesel generators, which at the time only have enough fuel to operate for ten days.

This sets off international alarms. The director general of the International Atomic Energy Agency (IAEA) warns that the latest attack could result in a Fukushima-like radiological release unless the agency can gain access to the facility and assure proper safety measures are being taken. Fighting near the plant, however, makes it dangerous to access. Nonetheless, in a repeat of 2022, the IAEA manages to send a team of inspectors to help "stabilize" the plant and avert any radiological release.

Unfortunately, the opposite situation unfolds. While the IAEA staff are en route, the Russian military bombs and disables the main paved corridor to the plant. One of the Russian strikes glances an IAEA vehicle, injuring an inspector. Russia indignantly denies any responsibility but joins Ukraine and the IAEA in calling for an emergency meeting of the UN Security Council.

NATO fears Russia will use its military garrison at Khmelnitsky to strike the surrounding areas with impunity. The U.S. team cautions that fighting near the plant could cause a radiological release and ultimately a meltdown.

Advisors to the U.S. Energy Department recommend that the U.S.-designed AP1000 and Russian-designed VVER spent fuel be transferred from the pools to dry cask storage. They also warn that further military assaults on Ukraine's nuclear plants could produce radiological releases that would force the evacuation of communities both within and beyond Ukraine's borders. Most NATO advisors, however, deem the immediate probability of a major radiological release to be low.

Russia then attacks some of the diesel fuel storage tanks at the Khmelnitsky plant, leaving the plant with only several days' worth of diesel fuel to run the emergency generators. Shortly after, Ukraine releases a video providing evidence of Russia's responsibility for the attacks.

Ukraine calls for the IAEA to mediate a shipment of diesel fuel to the plant's site to prevent a meltdown. The United States and its NATO allies support this request. The United States demands that Russia move its missiles out of the reactor site and create a demilitarized zone around the plant. Ukraine confirms the presence of U.S. Westinghouse personnel at the Khmelnitsky plant, but they are unable to leave the site.

NATO advises the United States that the long-term threat of a radiological release is growing and asks Ukraine and Russia to shut down the last operating reactor at the site. NATO believes that shutting the plant entirely is a reasonable request and suggests enlisting China and the U.N. Security Council to pressure Russia into doing so. In response to these requests, Russia puts the reactor on normal shutdown mode and demands that the United States de-escalate the situation by ceasing to send arms to Ukraine. When asked to ensure the security of the IAEA inspectors, Russia agrees.

Ukraine asks the United States and NATO to issue a joint press release stating that Russia's actions were an intentional effort to trigger a meltdown of one of the Khmelnitsky nuclear reactors, that Russia's attacks of the plant constitute a war crime, and that the allies view any event involving radiological dispersal as an intentional use of a "nuclear weapon" by Russia against Ukraine. Washington is hesitant to grant Ukraine's request and confers with key NATO members regarding its position. NATO consensus, however, was impossible to achieve. A key concern was that backing Ukraine's position would undermine any opportunity to negotiate with Russia.

Disappointed, Ukraine issues its statement independently and makes its own military plans to regain control of the plant. For this purpose, Kyiv asks the United States and NATO for precision missiles, drones, electronic warfare equipment, and satellite intelligence—to which they both agree. Ukraine also asks the United States and NATO for tools and transformers to restore external electrical power to the plant, as well as Western experts to monitor the plant's safety and the security of the plant's staff. The United States and NATO demur and instead suggest the IAEA assume this role and that U.N. peacekeepers and the Red Cross be placed in and around the nuclear plants.

Washington, then suggests a hedging strategy. First, it asks NATO to pressure Russia to demilitarize a zone around the plant. Second, it asks NATO for help positioning troops, diesel fuel, and emergency generators at the border with Ukraine. NATO agrees and positions counter-battery radar, counter-battery missiles, and troops near the Ukrainian border in Southern Poland. NATO also puts troops on a 24-48-hour recall alert system to allow them to quickly deploy where needed. It also puts its Combined Biological Radiological Nuclear (CBRN) forces on standby.

Several NATO members believe Russia's aggression against the Khmelnitsky plant constitutes a significant security risk. They raise this issue but defer to the United States on whether to invoke Article 5. Washington is not ready to do so.

Move Two:

Fighting in and around the Khmelnitsky plant continues. Russia warns Ukraine's reckless targeting of Khmelnitsky risks a Chernobyl-like radiological release.

Meanwhile, Ukraine's counteroffensive against Crimea gains ground. As Ukrainian forces close in on Sevastopol, Ukraine discovers that Russia intends to launch a false flag attack against a small research reactor at the Sevastopol National University of Nuclear Energy and Industry. Russia fires a missile damaging the reactor, prompting a limited, local release of radiation. Moscow immediately accuses Ukraine of having attacked the reactor with NATO-supplied weaponry and again demands that Romania and Poland close their borders to any further deliveries of U.S. and NATO-supplied offensive weapons. Moscow further warns that failure to seal the border will result in Russia taking "proportionate" action against Poland and Romania.

Before NATO or the U.N. can meet, several missiles explode in the parking lots of the U.S. NuScale reactor in Romania and of the Westinghouse AP1000 and Korean KEPCO APR-1400 reactors in Poland. Poland and Romania temporarily shut down these plants. Meanwhile, Ukrainian forces push Russian troops out of key positions in Crimea. Panicked, Moscow announces it will target Poland and Romania further unless they immediately stop all military equipment transfers into Ukraine. The United States, NATO, and Ukraine try to confirm Russia's culpability for the strikes in Poland and Romania as a majority of NATO members are unwilling to attack Russian forces until there is proof of an imminent attack.

While NATO awaits definitive intelligence, Ukrainian technicians at the Khmelnitsky nuclear plant worry that they may soon run out of diesel fuel needed to run the emergency electrical generators. To replenish their diesel stocks, Ukraine asks NATO to deliver fuel to Khmelnitsky. The United States, concerned about U.S. Westinghouse employees unable to leave the Khmelnitsky plant, plans to use the extraction of its personnel as an opening to also bring fuel to the generators.

Ukraine again calls for the United States and other NATO member countries to condemn Russia's attack against nuclear power reactors as a war crime under Protocol 1 of the Geneva Convention and insists it constitutes a first use of nuclear weapons. Shortly thereafter, overhead surveillance confirms that the missiles that struck Romania and Poland were fired from Russian-occupied Crimea. European NATO members, sensing Washington's general reluctance to invoke Article 5, are uncomfortable doing so but fear Poland, who is closest to Ukraine geographically and politically, might now strike Russia unilaterally. They want to invoke Article 5 to prevent this and the expansion of the war that might follow if Russia is struck.

Ukrainian leaders, meanwhile, feel NATO has abandoned them. They begin planning a retaliatory strike against Russia. Initially, some Ukrainian officials privately advocate striking Russian nuclear reactors, natural gas infrastructure, and urban and political centers. They drop this suggestion, though, for two reasons. First, it is unclear how hitting these targets will bring victory. (Ukraine had too few long-range strike systems to take this expansive target set on.) Second, striking them would undermine Ukraine's position that striking power reactors is a war crime. For these reasons, any Ukrainian strike against Russia is limited to military targets.

Farther to the west, Ukraine continues its plans to retake the Khmelnitsky plant. As fighting intensifies near the plant, all power to the plant is lost when unidentified missiles hit these reactors' dedicated emergency diesel generators. Within hours, battery backup power for the reactors' instruments is also lost.

Fire trucks are not available to provide emergency cooling. Ukrainian experts have no access to the plant, but they assess there is now only enough convective cooling to prevent meltdowns for a day or so at the VVER reactors and perhaps a bit more for the AP1000 units. As for the spent fuel pools, these should remain stable for at least a week.

The United States moves backup power generators and diesel fuel to Ukraine's border. U.S. troops accompany these shipments. Russia warns that if NATO troops enter Ukraine, it would constitute an act of war against Russia. The U.S. troops and power backup equipment do not cross the border. The U.S. plan to rescue the Westinghouse employees also falls by the wayside as the crisis escalates, leaving U.S. citizens trapped at Khmelnitsky.

After three days, meltdowns occur within all the reactors. However, at one of the two the AP1000 reactors, things spin even further out of control. Not only does the automatic, passive water-cooling reserve system runs out and radioactive steam and hydrogen build up; but because there is no external electricity flowing into the plant, the AP1000 reactors' electric hydrogen igniters cannot be used to "burn" off the hydrogen. An accidental spark at one of the AP1000 units prompts an explosion. This produces a major breach of the reactor's concrete containment and a significant atmospheric radiological release. Ukraine calls for a ceasefire to allow evacuation. Russia does not respond to this request.

The AP1000 reactor's radiological release upsets Ukraine's military plans to retake the Khmelnitsky plant. Frustrated, Ukraine decides to retaliate against Russia with a major long-range missile strike against Russian Black Sea fleet bases, including the base at Novorossiysk, well within Russian territory. Ukraine's objective is to make a "proportionate" military strike against valued Russian military bases.

At the Khmelnitsky plant, shifting winds push the plant's leaking radiation towards Poland, prompting some officials to recommend evacuating Rzeszow and Lubin. Ultimately, this course is rejected and instead Polish authorities ask the citizens to stay-at-home. NATO also considers delivering air defense systems, including Patriot and THAAD units, to Ukraine. These would protect its nuclear plants and act as a first-line defense for the rest of Eastern Europe. Ukraine, though, is unaware of these NATO deliberations. Ukraine, still believing NATO has abandoned it, asks Poland to conduct joint military operations against Russia.

Russia learns about NATO's plan to send air and missile defenses into Ukraine and warns that, if these systems cross the border, they will become a legitimate military target. At this time, unidentified missiles hit diesel generators at the Polish plant and at the Romanian CANDU reactors. Russia claims again that this is Ukraine's doing. NATO, including the U.S. team, decide to go ahead with the deployment of air defense units to Ukraine, despite Russian warnings. NATO and the U.S. team also deploy air and missile defenses, fire trucks, emergency electric generators, and diesel fuel stocks at all the nuclear plants in Poland and Romania. Washington forward-bases B-21 bombers to Poland and Romania to deter further Russian aggression, invokes Article 4, and asks NATO to invoke Article 5.

From this point on, events unfold rapidly. The players have difficulty managing what ensues.

Ukraine asks NATO for additional intelligence, missiles, and drones so Ukraine can strike the Russian staging bases that attacked Poland and Romania. Some frontline NATO countries are sympathetic, but the United States and several legacy NATO members refuse to support such attacks.

Meanwhile, NATO-provided air and missile defense systems cross the border into Ukraine. In response, the Russians target them. Russia hits most of these air and missile defense systems after they cross the border. Russia hits one, however, while it technically is still in NATO territory. Further to the south, the radiation spewing from the Khmelnitsky plant forces Russian troops to begin to evacuate the site. At this point, Ukrainian forces decide to move in to open up a corridor for fire trucks to reach the reactors. However, their progress is stalled by Russian forces.

Eager to take action, even without U.S. or NATO support, Ukraine then launches a missile strike against the Russian strategic bomber airbase at Engels, which intelligence shows is a source of many airstrikes against Ukraine. Ukrainian missiles disable the air base's runways, fuel farms, and long-range bombers.

Per its previous warning that any strike against its territory would constitute cause to place its nuclear forces on high alert, Russia does so. The United States, United Kingdom, and France respond in kind and NATO formally invokes Article 5 but does not take immediate military action. Instead, the United States proposes providing support to a Ukrainian offensive against Crimea, leaving open the option of direct U.S. military involvement if that offensive is unsuccessful.

Game play ends at this point.

Arresting Nuclear Adventurism: China, Article VI, and the NPT

Executive Summary:

Given the current crisis in Ukraine, it's tempting to consider focusing on Chinese compliance with the Nuclear Nonproliferation Treaty (NPT) to be an academic indulgence. Giving into this inclination, however, would be a mistake. As dangerous as Russia currently is, China will be more threatening in the long run. As we are learning with Russia's violation of the 1994 Budapest Memorandum, enforcing binding understandings is critical lest violators run roughshod over law and good order. This is true with Russia's behavior in Ukraine. It is no less so with China's nuclear weapons buildup and its repeated refusal to join in good faith negotiations to limit its nuclear weapons activities, which is required by Article VI of the NPT.

This buildup and refusal clearly fly in the face of China's legal NPT obligations. The question is what might bring Beijing back into compliance. To get the answers, NPEC held a battery of workshops last fall, followed by a week-long diplomatic simulation. The game participants included U.S., Japanese, and Australian former and current officials and staff as well as outside experts.

The group concluded that Beijing is unlikely to comply willingly with the NPT anytime soon, but that U.S. and international security would still be best served by spotlighting Beijing's nuclear adventurism and suggesting diplomatic off-ramps to arrest its nuclear buildup.

What this entails can best be summarized by the group's four key findings:

1. **China's nuclear weapons buildup and its refusal to negotiate in good faith to limit its nuclear arsenal constitutes a clear NPT compliance concern**. Article VI of the NPT stipulates that each of the parties to the treaty "undertakes to pursue negotiations in good faith on effective measures relating to cessation of the nuclear arms race at an early date, and on a treaty on general and complete disarmament under strict and effective international control." It has specific, enforceable meaning. China's unwillingness to join in nuclear arms reduction talks with Russia and the United States until American and Russian arsenals come down to Chinese levels as China more than doubles its own nuclear numbers constitutes a not-so-subtle evasion of Article VI. Beijing's prevarication is compounded by evidence of China's nuclear buildup (e.g., the construction of hundreds of new missile silos, Beijing's crash "civilian" plutonium production efforts to expand its nuclear stockpile, the suspicious activities China is conducting at its nuclear test sites, etc.). China gives token support to gridlocked talks to ban "weapons" in space and discussions on a Fissile Material Cut-off Treaty (FMCT). It also participates in a dedicated forum that brings together the five nuclear weapon states of the U.N. Security Council. It supports disarmament resolutions at the U.N. and professes a no first use policy. None of this, however, releases China from its Article VI requirement to negotiate in good faith with the United States and other NPT nuclear weapons states to limit nuclear arms. At best, referring to these forums and proposals constitutes diplomatic smoke designed to deflect requests China join in specific negotiations to reduce nuclear numbers. Given

bureaucratic conservatism within the U.S. Department of State, finding China in clear, legal violation of Article VI is unlikely. Finding that its nuclear behavior raises NPT compliance concerns, however, is bureaucratically feasible.

2. **Pressuring Beijing to uphold its NPT obligations may afford strategic benefits to Washington and its allies, but these are unlikely to be realized unless the United States and other NPT nuclear weapons states back effective nuclear restraints**. None of the game's participants believed pressuring China to enter in earnest nuclear talks would be easy. Yet, all believed taking China to task regarding its Article VI obligations could help increase the diplomatic costs of China expanding its nuclear arsenal, maintain the integrity of the NPT, and strengthen U.S.-allied ties and U.S. relations with nonaligned states. None of this is likely to obtain, however, if Washington and its nuclear-armed allies are seen to be trying to gain greater nuclear superiority over China and have no serious nuclear limitation proposals to offer. In this case, nonaligned states would continue to attack the United States, the UK, and France at the United Nations and in other multilateral disarmament forums (e.g., in Geneva, New York, and Vienna). If, on the other hand, Washington and its nuclear armed allies promoted new, effective nuclear controls (limits that might be made contingent on China or Russia also agreeing to comply in some fashion), the group thought many states, including nonaligned nations, would likely support efforts to get China to reciprocate and uphold its Article VI obligations. All of these observations are salient to what America and like-minded states might say at this August's NPT Review Conference; the annual sessions of the U.N. First

Committee; the sessions of the Conference on Disarmament; and at the meetings of the IAEA Board of Governors this September and November.

3. **A successful NPT campaign regarding China's noncompliance, would require Washington to shed normal modes of operation**. At a minimum, Washington would have to share much more sensitive intelligence on Chinese nuclear activities than it has with allies as well as nonaligned nations. It would have to declassify much of this information to convince the world that China's nuclear buildup is both real and threatening. Beyond that, Washington would need to promote talk on nuclear limits to uphold its Article VI obligations. Ideally, this would entail proposing nuclear control positions with other nuclear armed states — e.g., by restricting plutonium production for any purpose, establishing nuclear hotlines, clarifying the terms of the Comprehensive Test Ban Treaty (perhaps at a P-5 meeting), adopting new nuclear transparency measures (including more reporting to the UN, the IAEA, and IAEA safeguarding of Chinese and nuclear states' most militarily useful "civilian" nuclear facilities), freezing or reducing strategic nuclear weapon numbers, etc. Care would have to be taken, however, not to rush to get to yes embracing positions — e.g., jettisoning missile defenses or of America's extended nuclear deterrence policies — that might fray Washington's ties with its closest allies. Securing the active support of our allies and that of non-nuclear weapon states (e.g., members of the Nonproliferation and Disarmament Initiative or NPDI, the Stockholm Initiative, etc.), would be critical to any effective Article VI China compliance campaign. Washington also would have to take an even more active role than it currently does in international nuclear control

forums in Geneva (at the Conference on Disarmament), Vienna (at the International Atomic Energy Agency), and New York (at the U.N. First Committee, U.N. Disarmament Committee, and the NPT Preparatory Committee and Review Conferences).

4. **Military leverage will be important to move China on Article VI, but, in the short and long-term, the highest military leverage may not be nuclear**. Trying to leverage China's strategic nuclear buildup by keeping up or getting further ahead quantitatively on a nuclear-warhead-for-nuclear-warhead or nuclear-missile-for-nuclear-missile basis will hardly play to America's advantage. For the next decade or more, China will be able to produce new nuclear warheads and missile delivery systems much more cheaply and quickly than the United States. Rather than enter into a quantitative nuclear arms race with China, the United States should make it more difficult for China to target America's nuclear weapons. This should be done by making U.S. nuclear systems more difficult to locate, disable, and destroy. There are several ways to do this. The U.S. can make its ballistic missile submarines stealthier and its land-based missiles mobile. Virtually proliferating the number of possible air-based nuclear delivery systems by basing a few nuclear weapons on platforms that are numerous and making U.S. command, communications, control, and surveillance systems far more secure and survivable would also help. Ultimately, it is unclear how important nuclear weapons will be to providing strategic deterrence as more discriminate ways of disabling countries are developed in the coming decades (e.g., with high-precision, long-range unmanned conventional strike systems; cyber warfare and crypto technologies; unmanned surface and underwater naval warfare and sensors systems;

resilient space-based military systems etc.). Conversely, America and like-minded nations will hardly be able to isolate China regarding Beijing's nuclear buildup, if the United States, which currently has more nuclear weapons than China, chooses to acquire even more.

Wargame Format and Scenario:

The simulation was played in three moves: (1) a mock U.S. interagency meeting, (2) a mock Pacific summit of U.S. allies, and (3) a hot wash to discuss findings and develop strategy recommendations. Move one began in 2022 and participants engaged in a mock U.S. interagency meeting to develop a range of U.S. diplomatic, international, and allied coordinated actions for meaningful nuclear arms control negotiations with China. The purpose of these negotiations was to spotlight China's nuclear expansion and encourage China to adopt forms of nuclear restraint.

Move two simulated a Pacific summit hosted by the United States with Japan and Australia to stress test the U.S. strategy with key U.S. allies. Participants debated the merits of America's case to take action regarding China's compliance with the NPT and what international forums to share concerns about China's nuclear activities. They also identified useful pressure points that could move the PRC to negotiate and what intelligence Washington needed to share with whom.

In move three, the group discussed the simulation and its key findings.

Move One – United States Interagency Principals Meeting

In move one, the U.S. Director of National Intelligence briefed principals at the interagency meeting on the latest U.S. intelligence regarding China's nuclear activities (see appendix for full U.S. intelligence briefing). China had refused to participate in U.S.–Russian nuclear talks, citing its refusal to engage until Russia and the U.S. come down to its level.

The United States had also received intelligence of suspected Chinese reprocessing of power reactor fuel and use of civil reactors to produce tritium. This reprocessing was in violation of China's peaceful end use agreements with the United States and possibly others, and in possible violation of the PRC's 1997 IAEA voluntary reporting understanding.

After Washington's fourth request that China join talks underway with Russia to "pursue negotiations in good faith on effective measures relating to the cessation of the nuclear arms race," China rejected the invitation and was suspected of short cycling and reprocessing spent fuel from its civilian power reactors including the U.S. AP 1000 reactors and Westinghouse-designed Chinese design variants. Several Senators raised concerns that China may have violated its promise not to reprocess U.S.-origin spent fuel without prior U.S. consent as well as previous concerns about nuclear testing. Washington also had evidence that China was continuing to assist Saudi Arabia in production of yellow cake.

U.S. intelligence had difficulty verifying Chinese reprocessing of power reactor fuel. However, multiple intelligence sources noted reprocessing activity and the movement of a significant number of tritium containers from a variety of power reactor sites to one of China's tritium extraction plants. Washington also identified an increase in activity at China's Lop Nur nuclear weapons test site. China blocked data from its international monitoring stations.

PLA commentators discussed the need to expand their nuclear arsenal. Meanwhile, PLA commentators also stated the need to field roughly 2,500 nuclear weapons to deter the United States. At the NPT Review Conference, Japan, which had not yet begun operation of its own reprocessing plant at Rokkasho, objected to China's suspect nuclear activities as being in violation of Article VI of the treaty. In response, the Chinese complained about Japan's planned operation of reprocessing and enrichment facilities at Rokkasho and its connivance in the Quad, AUKUS, and U.S.-Japan nuclear-related activities.

Given this background, principals met at a U.S. interagency meeting to develop a strategy for a possible Pacific summit. The group questioned what type of intelligence, in what volume should be shared with whom regarding the Chinese nuclear buildup. It agreed that the United States would need to share as much intelligence as possible, even though it risked leaking to the Russians.

The Secretary of Energy suggested that Washington ask the IAEA to inspect the most troubling facilities to verify the intelligence. The Vice President said that if the United States could show that China was diverting material from its civil nuclear sector to its military sector, our allies would follow our lead. The National Intelligence Officer for WMD said that this activity would be apparent upon inspection and convincing. Should China refuse IAEA inspections, that would indicate Beijing had something to hide and also help Washington build a broader coalition.

When questioned whether the United States would be in danger of a successful first strike by China, the Secretary of Defense assured the group that Washington would retain a survivable second strike. However, East Asian allies would be deeply concerned about the United States' ability to maintain nuclear security assurances. This would be greatly troubling to the Japanese and South Koreans, who might decide to pursue their own nuclear program.

The National Security Advisor questioned how practical it was to find China in violation of the NPT. The Secretary of State suggested a softer approach: Rather than try to prove China was in violation of the NPT — a hard sell — find Beijing's behavior to raise NPT compliance concerns. This softer sell would help consolidate general support to pressure China to restrain its nuclear activities and politically penalize them for not doing so.

The most effective case against the Chinese entailed getting non-traditional U.S. partners to speak up. These would include the Non-Aligned Movement (NAM), the Group of Latin American and Caribbean countries (GRULAC), the Africa Group, state parties to the Treaty on the Prohibition of Nuclear Weapons, and the Association of Southeast Asian Nations (ASEAN). A strong statement from NATO and non-traditional U.S. allies might even pressure Beijing to come to the negotiating table.

The Secretary of Energy warned pointing the finger at China could backfire. China could turn the tables on the United States, claiming Washington, not China is in violation of Article VI due to U.S. behavior and numbers of warheads. However, other principals pushed back, stating the United States' position is far more defensible than that of China. Washington has been acting in good faith to meet its NPT obligations by reducing its arsenal and engaging the Russians in arms negotiations. While there is nothing new in accusing the United States of violating Article VI, evidence of a massive Chinese build up and Beijing's refusal to come to the table for talks are quite recent developments that are more destabilizing.

The group agreed that future Nuclear Posture Review and the National Defense Strategy-related presentations should clarify how the United States interprets Article VI, explain how the United States has upheld its obligations to the treaty, and spell out what other nations need to do. Presenting this positive narrative at forums, such as the U.N. First Committee, would serve as a contrast to China's nuclear misbehavior. Whenever appropriate, Washington should call for resolutions reaffirming each nation's obligation to negotiate in good faith under Article VI.

The discussion then turned to asking for action at the U.N. or IAEA. If Washington had enough credible evidence of China diverting civil nuclear materials to weapons production, it could approach the IAEA director general and ask that he request inspections of suspected Chinese facilities. If China pushed back, that would be telling. The Director of National Intelligence agreed, but suggested it would be better for a U.S. ally to approach the IAEA, so not to appear that the United States was exerting undue influence over the IAEA.

The National Intelligence Officer for Weapons of Mass Destruction suggested that U.S. allies could act as well: Japan could agree to hold off opening the Rokkasho reprocessing plant, South Korea could delay start up reprocessing, and India might make an equivalent offer.

The conversation then shifted to how the U.S. Congress might legislate to exert pressure on China. The Vice President reminded the group of bipartisan legislation introduced by Senators Rubio and Markey in 2016 calling for a provision in the U.S. law that stated, "it is the sense of Congress, that the United States should encourage countries in East Asia to forego the commencement of new spent fuel reprocessing activities as part of a mutual effort to prevent the increased or expanded stockpiling of separated plutonium in the region." The National Intelligence Officer for WMD suggested that Congress could require the president to certify that China was in compliance with the NPT as a stipulation for continuing some action China might want the U.S. to take or to continue taking.

The group decided to call a Pacific Summit with Australia and Japan that would promote the following three ideas:

1. There should be two intelligence briefings. One would share our most sensitive information with the Five Eyes and our closest allies. Another would be less detailed for presentation to key nonaligned states.

2. Give the IAEA director general the first brief and ask him to request an inspection of key civil Chinese nuclear facilities.

3. Explore with Australia and Japan the practicality of pausing advanced reactor and key fuel cycle activities.

Separately, the group decided to encourage Congress to:

1. Build on the Rubio Markey 2016 legislation to produce a resolution calling on nations around the Pacific Rim to pause commercial deployment of fast reactors and spent fuel recycling.

2. Pass legislation requiring the President to certify that China is in compliance with the NPT as a condition for continued trade in certain sectors.

The Pacific Summit

The Director of National Intelligence shared the details of the U.S. intelligence briefing. In specific, the U.S. sought to 1) reaffirm the significance of the NPT and make noncompliance with the treaty a public issue; 2) ask the IAEA to inspect the Chinese facilities under question; and 3) ask Japan not to open its Rokkasho facility. The United States asked the allies for their opinions on the degree to which these actions would succeed in changing China's nuclear trajectory and what other actions could be taken.

The U.S. National Security Advisor underscored the need to call China out on its noncompliance with Article VI of the NPT, stating that if Beijing's intentions were good, then emphasizing Article VI would allow China to demonstrate its good faith. However, if the PRC's intentions were more sinister, Article VI could serve as rallying cry to encourage other nations to join the U.S. and allies in addressing the Chinese nuclear threat.

Both the Japanese and Australian delegations agreed that focusing on Article VI would strengthen support for the treaty. They also recommended pressing China to open its civil facilities to inspections, thereby broadening the remit of Article IV. Finally, they suggested taking China to task for proliferating nuclear weapons-related technology to North Korea, Pakistan, Saudi Arabia, and Iran.

Regarding America's vulnerability to counter charges of being in violation of Article VI, the Japanese Foreign Minister suggested Washington take a number of arms control initiatives. Among the ideas considered was clarifying the terms of the Comprehensive Nuclear Test Ban Treaty (what is a nuclear test). This could pressure China to formalize its moratorium on nuclear testing. The United States should renew negotiations on the Fissile Material Cut-off Treaty, asking China to support the spirit of the treaty by suspending production of fissile material for nuclear weapons.

Japan's Defense Minister suggested that the United States, Russia and China enter into a regional INF treaty with Japan and the Far East. The Australian government might enforce its bilateral agreement on the supply of uranium to China, insisting on the end-use assurance that uranium provided by Australia be used only for civilian purposes. To counter potential Chinese claims that the United States allowed India to use American uranium for military purposes, Washington could ask India to restrict its uranium to civil purposes as well.

The Japanese Foreign Minister noted that suspending work on Japan's planned reprocessing plant at Rokkasho would be difficult. Only if China and South Korea were asked not to operate their commercial reprocessing plants, might Japan freeze operations at Rokkasho.

The Summit then turned to the question of what Washington and its allies should be prepared to do if China came to the negotiating table. China would likely ask the United States to reduce its nuclear arsenal, which would require the Russians to do likewise. Beijing would also want the United States to limit its missile defenses.

Australian officials said the time was not right, geopolitically, to begin the denuclearization process and urged Washington not to step back from its nuclear commitments to its allies in the Pacific by reducing its nuclear arsenal. Should the United States fail to constrain China and accept mutual vulnerability or parity with China publicly, it would undermine U.S.-allied relations.

Participants considered, ultimately, what would happen in the region if the United States failed to maintain its relative nuclear superiority vis-a-vis China. In this case, South Korea, Japan, and Taiwan might get nuclear weapons of their own. Strong anti-nuclear sentiment in Japan and public support for the Treaty on the Prohibition of Nuclear Weapons would likely make it the last country in the region to acquire weapons.

South Korea, though, is another matter. Conservatives in South Korea have already discussed obtaining weapons or seeking American help to re-introduce U.S. tactical nuclear weapons in the country, and the South Korean public does not have the same apprehensions against nuclear weapons as the Japanese.

TPNW issue and suggested that the Nonproliferation and Disarmament Initiative (NPDI) – grouping of 12 cross regional countries, focused on pushing consensus outcomes at NPT (Japan and Australia started in 2010, Germany Netherlands, Nigeria, UAE, Mexico, Chile, Philippines, UAE, Turkey) – should make efforts to be more visible at the NPT conference.

The Australian Foreign Minister agreed, stating that the NPDI is already making progress in this area. Success depends on how the Article 6 issue is raised. Australia has come close to making this point publicly already, and its Ambassador to China said publicly at a conference in China that "it will be important for other nuclear weapon states, including China, to embrace these developments and – consistent with their obligations under Article VI of the NPT – commit to taking similar steps towards achieving meaningful disarmament. We encourage China to engage in Strategic Stability talks with the United States." The Australian Summit concluded with agreement to pursue the following action plan, which was subsequently confirmed in the game's concluding hot wash session:

<u>International Initiatives</u>

> The U.S., its allies, and supporters should spotlight China's noncompliance with Article VI of the NPT. While China might not immediately change its nuclear course, spotlighting its conduct would start a useful discussion. It would garner more support internationally to have Japan and Australia take the lead rather than Washington.

Encourage the Nonproliferation and Disarmament Initiative (NPDI consisting of Australia, Japan, Germany, Netherlands, Nigeria, UAE, Mexico, Chile, Philippines, UAE, Mexico, and Turkey) to raise noncompliance concerns about China and Article VI. They might build on the request Australia's ambassador to China recently made of Beijing to embrace the world's desire to reduce nuclear weapons "and – consistent with their obligations under Article VI of the NPT – commit to taking similar steps towards achieving meaningful disarmament."

Propose that all weapons states under the NPT open their civil nuclear facilities to IAEA inspectors. This proposal should be introduced by non-nuclear weapons states, such as Japan, Australia, Ireland, the UAE, Mexico or Canada. The United States should announce that it will allow its civil enrichment and any fast reactor related activities to be safeguarded by the IAEA. The United States should implore China to do the same.

Ask Australia to follow in the footsteps of Taiwan and the UAE to announce a timeout on any plans to enrich or reprocess.

<u>U.S. China Initiatives</u>

The United States should offer to consider agreements that would cap all NPT weapons states' nuclear arsenals at agreed levels.

The United States should propose that China engage in stability talks (regarding such ideas as hotlines), not only with Washington but with its key neighbors.

Allied Initiatives

The United States and Japan should consider attending the Treaty on the Prohibition of Nuclear Weapons conferences as an observer.

The P-5 should issue a joint statement reaffirming their continued adherence to a nuclear weapons test moratorium, which could politically bind China, and agree to further clarify what a nuclear test is.

Propose that all nations on the Pacific Rim including Australia, China, North and South Korea, Japan, and the United States take a timeout from civil recycling of plutonium.

IRAN LEAVING THE NPT: OUR NEXT HEADACHE

Executive Summary:

This report reflects what happened during a three-move diplomatic simulation NPEC hosted in August 2021 and in the game's three preparatory meetings. The simulation was set in the present day (summer 2021). Participants responded to a scenario in which Iran threatened to withdraw from the NPT, after which Israel released photos of Iran secretly manufacturing implosion devices. Participants assumed the roles of members of the U.S. National Security Council, key foreign officials, and members of the United Nations Security Council. In the first move, the U.S. President, responding to the crisis, called a National Security Council Principals Committee meeting to draft a U.S. resolution to the U.N. The U.S. Ambassador to the U.N. presented this resolution during a mock U.N. Security Council meeting in the second move. The resolution did not pass. In the third move, participants debated what strategies might ensure the future viability of the NPT and prevent Iran from breaking out and making a bomb.

The simulation produced five takeaways:

1. **The NPT's future enforcement will depend on Washington working with like-minded nations far more than it has ever done before.** The game's play demonstrated that Washington's authority to convince other members of the U.N. to enforce the NPT has declined as a result of its nuclear misadventures with Iraq and North Korea. Not only will the United States have problems convincing China and Russia to take action against Iran,

but it will also have difficulty doing so with a good number of friendly states. This disadvantage will persist for some time. As such, it is vital to gain general agreement on a common sense reading of the NPT. Fabricating nuclear warheads awaiting the insertion of nuclear explosive cores — are and should be understood to be clear violations of the treaty. Also, country-neutral rules should be in place that stipulate that states withdrawing from the NPT must clearly be in compliance and must still keep their civil nuclear materials and activities under IAEA safeguards.

2. **The IAEA is likely to become the first stop for any future NPT enforcement**. The game's play demonstrated the advantage of the United States and like-minded nations making a direct appeal to the IAEA director general to conduct special inspections to confirm evidence of a violation. In the future, this will likely be necessary before taking any major U.N. action. This will require the United States to share its most sensitive intelligence with key allies and the IAEA to convince the director general to secure corroborating evidence of possible NPT violations. To make enforcement of NPT violations more effective it would be helpful if both the IAEA and the U.N. had full-time staff to work such issues. Finally, to be ready to react properly to possible NPT and IAEA violations and withdrawals, the United States should encourage key U.S., allied, and international organizational staff to participate in simulations of such scenarios.

3. **Washington and other like-minded nations must resolve the growing tension between relying primarily on the national use of force against NPT violators versus applying broad diplomacy pressure.** States most interested in blocking Iran from getting the bomb (e.g.,

Israel) believe force must be used to achieve this goal and have little faith in actions of the U.N. or the IAEA to enforce the NPT. They favor having the United States conduct military strikes unilaterally or with a limited number of allies against NPT violators. For them, the NPT cannot survive unless violators clearly understand they will pay a high price, whether authorized by the U.N. or IAEA or not. Another view is that such authorization is critical to long-term, international support of the NPT and IAEA. Resolving the tension between these two views will be central to securing the NPT's survival. Getting this right is of no small moment: The NPT still remains the sole international legal justification for taking enforcement action against states that might try to acquire nuclear weapons.

4. **If Iran does get the bomb, sanctioning it economically would still be useful to stigmatize its behavior and strengthen future enforcement of the NPT.** Imposing significant economic sanctions until Iran comes back into compliance would be useful to keep Iran in a penalty box until and unless it comes back into compliance with the treaty. With Iran stigmatized, it would be easier to act earlier against lesser violators of the treaty. Stigmatizing bad behavior early would shore up adherence to the NPT far more than taking tougher action later.

5. **Israel's unacknowledged nuclear arsenal and its refusal to limit its nuclear weapons efforts puts America and like-minded nations at a disadvantage in enforcing the NPT in the Middle East and beyond.** Some believe asking Israel to limit its nuclear weapons programs in any way would imperil its security. Others believe it's absolutely essential to rid the Middle East of nuclear

weapons – a goal whose ultimate achievement is essential to Israel's survival. The NPT review conferences have repeatedly called for creation of a Middle East nuclear-weapons-free zone, and Iran has accused Israel of nuclear hypocrisy. On these matters, the United States has backed Israel and insisted that Israel does not have a nuclear arsenal. Although this approach has worked for many decades, it will be more and more difficult to sustain.

Simulation Format and Scenario:

Move one was a mock meeting of the U.S. National Security Council. The scenario was set in the present day, the summer of 2021. Throughout July 2021, U.S. intelligence agencies intercept communications between senior Iranian officials and Iran's Supreme Leader suggesting that Iran would follow through on repeated threats to leave the NPT unless its right to enrichment and reprocessing was recognized and all economic sanctions were lifted. On July 27th, U.S. intelligence agencies receive information suggesting the Supreme Leader had decided Iran's best option now was to build a nuclear bomb.

The next day, in consultation with their Israeli intelligence counterparts, CIA officials view Israeli photographic evidence that Iran is manufacturing nuclear implosion devices. U.S. intelligence agencies had previously determined that Iran assembled and tested a single device without highly enriched uranium (HEU), but the photographs the Israelis shared with the CIA were of the assembly of several devices, not one. On July 30th, The Israelis approach the U.S. Secretary of State with the photographic evidence and ask that the United States support a major Israeli strike against Iran's implosion device plant. In specific, the Israelis ask the Secretary to provide U.S. air refueling to Israeli strike aircraft on their way to

Iran. The Secretary refuses, stating that the United States will follow the rule of law and take diplomatic action first. The following day, the Israelis leak their photographic evidence to The New York Times, which runs the story with pictures late on the afternoon of Sunday, August 1st.

On the morning of August 2nd, the President holds a press conference and publicly states that, after conferring with the U.S. intelligence community, it is his belief Iran is in violation of the NPT. Within minutes, an Iranian foreign ministry press spokesman goes on TV rejecting the accusation, claiming the Israeli photos are fakes. This spokesman, then announces that Iran has begun the clock on withdrawing from the NPT. The spokesman notes that the clock would continue to run until the White House promises to block further Israeli military strikes against Iran and recognizes Iran's entire nuclear program as being completely peaceful. That afternoon, the U.S. Senate passes a resolution calling on the Administration to take immediate action to enforce the NPT. The President calls for a National Security Council Principals meeting to draft a U.S. resolution to the U.N. Security Council. This is where move one begins.

Move One – U.S. National Security Council Principals Meeting

Participants are presented with three options developed by the National Security Council deputies. The first is a U.S.-led enforcement action, including both economic sanctions and military strikes against Iran. The second option is to work with China, Russia, and the EU to get Iran to immediately resume participation in the Joint Comprehensive Plan of Action (JCPOA). The third option is to work cooperatively with the EU and key Pacific allies to draft a U.N. Security Council resolution that would strengthen economic sanctions against Iran. These sanctions would increase penalties every time Iran took an escalatory step. Participants debated a range of policies based on these options, and most agreed that a second NPT enforcement failure, after North Korea's withdrawal in 2003, could seriously, if not permanently, weaken the current nuclear nonproliferation regime.

A combination of Israeli and U.S. military strikes was discussed to delay, not to eliminate, the Iranian nuclear program for a period of years. Participants recognized that even if Israel chose to take unilateral military action, it would fail to eliminate the Iranian nuclear program and could draw the United States into a war to "finish the job" in Iran. Participants noted America's lack of political will and domestic support for entering a war and that the United States might have to live with a nuclear Iran. This resulted in the participants adopting a compound approach: Washington would make it clear that Israel should not act unilaterally but the United States would threaten military force to signal to Iran and other states that NPT violations or withdrawal would have serious consequences.

Most of the game participants recommended that the United States first pursue a diplomatic effort to validate U.S. and Israeli nuclear intelligence on Iran's manufacture of nuclear implosion devices. The U.S. could go directly to the IAEA director general to initiate immediate IAEA special inspections of the Iranian nuclear sites. If the United States could persuade the IAEA director general to act, Washington would not need Chinese or Russian support. The participants discovered that this should be done before proposing a resolution to the U.N. Security Council as the IAEA would be unlikely to agree to special inspections should a U.N. resolution on this topic fail. As such, the IAEA should be the first step in validating and enforcing NPT violations.

One concern raised by the participants is that the IAEA continues to favor the position that the agency is equipped and designed to account for nuclear materials, not discovering illicit nonnuclear weapons-related activities or determining if the NPT has been violated. Yet, it is clear that acquiring or developing implosion devices violates the NPT and the IAEA is better equipped than the U.N. to call out such violations.

Given the possibility that Iran was preparing nuclear implosion devices, the group drafted a U.N. resolution that called for the U.N. Security Council, acting under Article 41 of Chapter VII of the U.N Charter, to demand Iran cease work on any implosion devices and suspend its withdrawal from the NPT. The resolution further authorized the IAEA to inspect any suspect Iranian sites within a matter of days and to report on its findings. Should Iran complete its withdrawal from the NPT, the resolution stated that Iran must be prohibited from transferring, controlling, or manufacturing nuclear weapons or explosive devices. (See appendix for the full text of the U.N. resolution.)

Move Two – United Nations Security Council Meeting and Its Repercussions

In move two, participants assumed the roles on the United Nations Security Council, including the five permanent members, a select group of non-permanent members, and interested states including Iran, Israel, the UAE, and Germany. The participants debated the proposed U.S. resolution and voted. Following the vote, the U.S. National Security Council met to discuss its response to the outcome of the U.N. Security Council meeting.

At the start of this move, the U.S. Ambassador presented the proposed U.S. resolution to the U.N. Security Council. Member states' positions were predictable – U.S. allies supported inspections, while the PRC and Russia created roadblocks. Iran was concerned that the U.S. and Israel would use intelligence obtained during inspections to guide future military attacks. It also noted that the resolution failed to call for all states to join the NPT, acknowledge that Israel is a nuclear weapons state that has threatened Iran, or state that the current crisis must be resolved peacefully without military force.

There was consensus about the need for inspections to validate U.S. intelligence claims. America's nuclear misadventures with Iraq and North Korea undermined its credibility at the U.N. in identifying and enforcing NPT violations. Even allies, such as France, called for more credible evidence of Iran's violation before agreeing to the U.S. resolution. Most nations, including U.S. allies, Russia, and Iran, agreed that an amendment requiring the United States to rejoin the Joint Comprehensive Plan of Action (JCPOA) was needed. Ultimately, Russia and China vetoed the U.S. resolution. The PRC and Russia refused to support any resolution unless it reaffirmed Iran's sovereignty, prohibited military intervention, and required the U.S. to submit its intelligence sources to deep scrutiny and verification.

Following the U.N. Security Council vote, participants were asked to return to their U.S. National Security Council roles to develop a strategy in the wake of the resolution's defeat. America's next step was to approach the IAEA to request special inspections, separate from the U.N. The idea here was to exploit Russia's and China's inability to veto such a request. Assuming that the Iranians would allow inspections, that the inspectors found credible evidence, and that the United States could convince the Israelis to hold their fire, the United States could go back to the U.N. with sufficient proof to pass a resolution.

If the international community was unable to reach agreement on making Iran open up to inspections, the U.S. planned to impose multinational economic sanctions until the Iranians did open up. At the same time, the United States and its allies agreed to move forces to the region with the intent and ability to enforce the sanctions.

Finally, the United States tried to convince Israel to take a diplomatic approach. To do so, the U.S. Secretary of Energy agreed to brief Israel on U.S. estimates of when Iran could acquire five weapons (the number of weapons the Israelis themselves have argued Iran must get before it could pose a military threat). The Energy Secretary gave the brief, which pegged the acquisition date near the end of 2022 and argued that this gave adequate time to allow diplomacy to work. Washington made it clear to Israel that if it chose to strike Iran, it would be doing so unilaterally and could not assume U.S. support. America insisted that Israel work through the IAEA and share any and all of its evidence with Vienna, as would the United States. In the unlikely event that Israel was attacked conventionally or by nuclear weapons, the United States made it clear that it would come to its defense.

Move Three – U.S. Strategy and Hotwash

The purpose of the third move was to discuss the key takeaways from the simulation by focusing on a series of questions regarding the strengths and weaknesses of the proposed U.S. strategy and future value of the NPT. But first, participants resumed their roles on the U.S. National Security Council to refine Washington's strategy for addressing Iran's NPT withdrawal.

Participants struggled to address an underlying dilemma – was it more important to immediately prevent Iran from getting the bomb or to uphold the NPT? The unspoken assumption was that if Iran succeeded in getting a nuclear weapon, the NPT would fail. Therefore, some participants argued, saving the NPT was secondary, and the United States should use any and all means, including U.S. and allied military efforts that lacked international authorization, first to block Iran from the bomb.

On the other hand, such a military approach could undermine future international support to enforce the NPT, the sole international legal basis for taking military and diplomatic action against states attempting to acquire nuclear weapons. If the primary goal is to protect the NPT to prevent other states from acquiring nuclear weapons, taking a diplomatic approach with Iran would be essential. This, in turn, would require making early, direct contact with Russia and China. Although some participants called for contacting Russia, China, and Iran, the participants skipped taking this course in the game's play.

Regardless of how close Iran or any other nation might get to acquiring a nuclear weapon, most game participants thought it was critical to impose stringent sanctions against any and all NPT violators as early as possible and to maintain them until these states ended their weapons programs and gave up any weapons they may have made (think South Africa). This would send a message to proliferators that any nation that builds or acquires nuclear weapons will suffer draconian sanctions unless and until they come back into compliance with the NPT.

Participants also emphasized the importance of first approaching the IAEA and getting it to verify any NPT violations-related intelligence, as IAEA inspections would provide essential groundwork for going to the U.N. Security Council. Making sure both the IAEA and the U.N. were properly staffed and organized to take needed NPT violation-related actions was discussed at an earlier game preparatory meeting. There is reason to believe neither agency is so organized.

In any case, if Iran stiff-armed IAEA inspectors or the inspections uncovered missing nuclear materials, this would provide an opportunity to approach the U.N. Security Council for a resolution. Approaching the U.N. Security Council first without sufficient, internationally validated evidence, would not only be risky politically, but give the Iranians more time to dismantle or move suspect materials or facilities to evade detection.

Because there is no one-size enforcement method to fit all scenarios, some in the game's preparatory meetings suggested having U.N., IAEA, and national officials game various NPT withdrawal and IAEA violations scenarios.

The game discussion then turned to the simulation findings, and participants answered three major questions:

1. *Do we understand Iran's motives and to the extent that we do, are any of them legitimate?*

Some participants believed Iran's motivation for developing nuclear weapons is primarily defensive. Iran's leadership aims to protect its republic and ideology. Not only has Iran watched the U.S. invade its neighbors, but it fears a Western cultural invasion will undermine the Islamic regime in Iran. Having a nuclear weapon, the participants noted, demonstrates to not only the West, but Iran's own people and neighbors, that the Islamic republic is strong and steadfast. Nuclear weapons provide Iran with a deterrent, and would help diminish the security divide between the United States and Israel. On the other hand, other participants thought Iran may only be seeking nuclear latency as a point of leverage to extract concessions from the United States in terms of sanctions relief.

This lack of clarity regarding Iran's motives highlights the need to have a way to be in "touch" with Tehran routinely and in crises and, more important, to upgrade our intelligence resources there.

2. *What are the risks and benefits of ignoring the NPT in dealing with the Iranian case? What legal alternatives might there be to the NPT?*

The NPT is the sole, international legal barrier to states that lack nuclear weapons from getting them. The treaty's vitality is itself a key metric for gauging the international norm against acquiring nuclear arms. For nuclear states that are not members of the NPT, the treaty is often viewed as an impediment to them taking military action alone against would-be bomb makers. The treaty is most important legal and diplomatic impediment to nations acquiring the bomb and can be used by the U.N. Security Council to authorize penalties, including military action, against violators.

The treaty, however, is only one of many levers against nuclear proliferation. States can stigmatize bad behavior without the clear authorization of the U.N. or the IAEA and do so even when that behavior cannot be prevented. In the case of friendly countries, such as South Korea, the specter of losing military, commercial, and political ties with the United States is still likely strong enough to keep it from getting the bomb. In Iran's case, economic, and trade sanctions may be the best option to stigmatize Iranian nuclear misbehavior. Although the game's play suggests that it might not be enough ultimately to deter Iran from going nuclear.

That said, ignoring what the NPT demands increases the risks of a nuclear arms race in the Middle East. Certainly, if Iran is permitted to remain in the treaty after it acquires nuclear weapons or is not penalized for violating the treaty, other nations in and outside of the region will strive to match Iran's capabilities. This is an important reason to impose severe additional international economic and trade sanctions on Iran if it should ever get the bomb or try to leave after violating it.

3. *Is there something the United States should be promoting to deter any country from threatening to withdraw from the NPT?*

Ultimately, nuclear weapons proliferation in the Middle East cannot be tackled without addressing Israel's nuclear weapons program. Nuclear deterrence was not Israel's original justification for developing nuclear weapons. In the 1950s, Israeli leaders saw the bomb as a failsafe should Israel lose a conventional war. For these historical reasons (and the Holocaust), Israel is resistant to dismantling its nuclear arsenal. However, some in this game noted that Israel's security may depend on none of its neighbors getting nuclear weapons and that this goal is best achieved by making the Middle East a nuclear-weapon-free zone. If so, Israel's ultimate security could require it to support some form of overt, diplomatic nuclear restraint (e.g., ending its own enrichment of uranium, placing some additional portion of its nuclear activities under international inspections, etc.). So far, this tension has not been addressed, especially in public. Whatever the merits of either side, the case for continuing relative official silence on these matters ought to be reevaluated.

Another recommendation the game's play supports is the desirability of strengthening the common sense view that a nonweapons state's acquisition or development of the key nonnuclear components of a nuclear weapon constitutes a clear violation of the NPT. Currently, these waters are muddied by the way many view the IAEA's original charter to verify and account for nuclear material inventories. The IAEA still reads this inspection mandate quite narrowly. As the agency's director general noted in 2006, "absent some nexus to nuclear material, the Agency's legal authority to pursue verification of possible nuclear weapons-related activity is limited." In practice, as well in law, though, the IAEA, can and has inspected even when nuclear materials are not present.[276] One idea discussed in one of the game's preparatory meetings was to create dedicated NPT compliance staffs both at the U.N. and the IAEA and to give the IAEA much more explicit authorization to inspect for nonnuclear weapons-related activities when appropriate.

Yet another recommendation that was discussed at length in the game's first preparatory meeting is having the United States and other like-minded nations support U.N. adoption of country-neutral rules to deter states from violating or withdrawing from the NPT. The most complete treatment of this option is that of Pierre Goldschmidt, former director general of the IAEA for safeguards. In his NPEC report, "Dealing Preventatively with NPT Withdrawals," Mr. Goldschmidt calls for the United States and like-minded states to back a U.N. resolution that would require states that threaten to withdraw from the NPT to keep all of their nuclear materials and facilities under IAEA safeguards (currently, states that withdraw can end such inspections) and to suspend enriching and reprocessing until the IAEA can certify that all special material they produce has been fully accounted for. He also calls for having the U.N. stipulate in advance that if any state withdraws from the NPT, the U.N. will consider this to be "a major threat to international peace and security" under Article 39 of the Charter of the United Nations, which, in turn, would allow the U.N. to sanction the withdrawing state. Finally, he would have the permanent five members of the U.N. Security Council promise not to veto the sanctioning of any withdrawing state if they were the sole state to do so. Several of the game participants noted that the United States had backed such proposals in the early 2000s but

---

276. See George Bunn, "Nuclear Safeguard: How Far Can Inspectors Go?" IAEA Bulletin, March 2007, available at https://www.iaea.org/sites/default/files/publications/magazines/bulletin/bull48-2/48203494955.pdf.

were unable to gain support for their passage. The outcome of the Iran simulation, however, strongly suggests that yet another attempt might be in order. In back briefing the game's results to U.S., foreign, and international officials, one other suggestion was made: Encourage the IAEA to game how it might deal with future IAEA violators and states trying to withdraw from the NPT. To date, no such simulations have been attempted or promoted. Given the results of this simulation, however, states, such as Saudi Arabia, the UAE, and other NPT members of the Abraham Accords might see merit in supporting and participating in such efforts.

Made in the USA
Monee, IL
04 November 2026

67386086R10114